DATE DUE

OCT 31 1979			
OCT 23 1979			
2-2-88	ILL		
5517780			

PROBABILITIES AND POTENTIAL

NORTH-HOLLAND
MATHEMATICS STUDIES 29

Probabilities and Potential

CLAUDE DELLACHERIE

Institut de Mathématique
Université Louis-Pasteur, Strasbourg

PAUL-ANDRÉ MEYER

Directeur de recherches
Centre National de la Recherche Scientifique

HERMANN, PUBLISHERS IN ART AND SCIENCE

1978

NORTH-HOLLAND PUBLISHING COMPANY - AMSTERDAM - NEW YORK - OXFORD

Hermann ISBN : 2 7056 5857 2
North-Holland ISBN : 0 7204 0701 x

Translation of : Probabilités et potentiel © 1975, Hermann, 293 rue Lecourbe, 75015 Paris. France

PUBLISHERS
HERMANN, PARIS
NORTH-HOLLAND PUBLISHING COMPANY
AMSTERDAM • NEW YORK • OXFORD

SOLE DISTRIBUTORS FOR THE U.S.A. AND CANADA :
ELSEVIER NORTH-HOLLAND, INC.
52 VANDERBILT AVENUE, NEW YORK, N.Y. 10017

Library of Congress Cataloging in Publication Data :

Dellacherie, Claude.
Probabilities and potential.

(North-Holland mathematics studies; 29)
Translation of Probabilités et potentiel.
Ed. of 1966 by P.-A. Meyer
Bibliography: p
Includes indexes.
1. Probabilities. 2. Measure theory. 3. Potential, Theory of. I. Meyer, Paul André, joint author. II. Meyer, Paul André. Probabilités et potentiel. III. Title.
QA273.D3713 519.2 77-26865
ISBN 0-7204-0701-X

PRINTED IN FRANCE

Contents

CHAPTER 0. NOTATION 1

CHAPTER I. MEASURABLE SPACES 7

σ-fields and random variables 7

Definition of σ-fields (no.1). Random variables (nos. 2 to 4). σ-fields generated by subsets, etc... (nos. 5 to 7). Product σ-fields (no. 8). Atoms, separable σ-fields (nos. 9 to 12).

Real-valued random variables 11

First properties (nos. 13 to 18). The monotone class theorem (nos. 19 to 24).

CHAPTER II. PROBABILITY LAWS AND MATHEMATICAL EXPECTATIONS 17

Summary of integration theory 17

Probability laws (nos. 1 to 4). Expectations, Lebesgue's theorem, etc... (nos. 5 to 9). Convergence of random variables (no. 10). Image laws (nos. 11 to 12). Integration of laws, Fubini's Theorem (nos. 13 to 16).

Supplement on integration 21

Uniform integrability (nos. 17 to 22). Vitali-Hahn-Saks Theorem (no. 23). Weak compactness, Dunford-Pettis Compactness Theorem (nos. 24 to 26). Rapid filters (nos. 27 to 29).

Completion, independence, conditioning 30

Internally negligible sets (nos. 30 to 32). Independence (nos. 33 to 35). Conditional expectations (nos. 36 to 40). List of their properties (nos 41 to 42). Conditional independence (nos. 43 to 45).

CHAPTER III. COMPLEMENTS TO MEASURE THEORY 39

Analytic sets 39

Pavings (no.1) Compact pavings (nos. 2 to 6). Analytic sets and closure properties (nos. 7 to 13). The separation theorem (no. 14). Souslin measurable spaces, etc... (nos. 15 to 17). Direct images (nos. 18 to 20). The Souslin-

Lusin theorem (nos. 21 to 23). Blackwell spaces (nos. 24 to 26).

Capacities

Choquet capacities (no. 27). Choquet's theorem (nos. 28 to 29). Construction of capacities (nos. 30 to 32). Applications : measurability of analytic sets (no. 33), Caracthéodory's theorem (no. 34) and Daniell's theorem (no. 35); measures on compact spaces (no. 36) and Lusin spaces (nos. 37 to 38). Left-continuous (nos. 39 to 40) and right-continuous (nos. 41 to 42) capacities. Another proof of the separation theorem (no. 43). Measurability of debuts and cross section theorem (nos. 44 to 45).

Bounded Radon measures

Radon measures (nos. 46 to 47). Filtering families of l.s.c. or u.s.c. functions (nos. 48 to 50). Inverse limits : Kolmogorov's theorem (nos. 51 to 52) and Prokhorov's theorem (no. 53). Strict convergence (nos. 54 to 58). Prokhorov's compactness criterion (no. 59). The space of probability laws (nos. 60 to 62). Lindelöf spaces (nos. 63 to 66). Non-metrizable Souslin and Lusin spaces (nos. 67 to 69). Desintegration of measures (nos. 70 to 74).

(Nos 75 to 86 appear in the appendix, see below).

CHAPTER IV. STOCHASTIC PROCESSES 8

General properties of processes 8

Processes (no. 1). Philosophy (nos. 2 to 5). Standard modifications and indistinguishable processes (nos. 6 to 8). Time laws canonical processes (nos. 9 to 10). Filtrations, adapted processes and philosophy (nos. 11 to 13). Progressivity (nos. 14 to 15).

Regularity of paths 89

Notation (no. 16). Processes on a countable dense set (nos. 17 to 19). Upcrossings and downcrossings, applications (nos. 20 to 23). Separable processes (nos. 24 to 30). Random closed sets (nos. 31 to 32). Progressivity of certain processes (nos. 33 to 34). Almost-equivalence (no. 35). Essential topology (nos. 36 to 39). Pseudo-paths (nos. 40 to 46).

Optional and predictable times 11

Definitions concerning filtrations (nos. 47 to 48). Stopping times (no. 49). Debuts (nos. 50 to 51). σ-fields associated to stopping times (nos. 52 to 54). Properties (nos. 55 to 59). Stochastic intervals (no. 60). Optinal and predictable σ-fields (nos. 61 to 63). Properties (nos. 64 to 68). Predictable times (nos. 69 to 74). Sequences foretelling a predictable time (nos. 75 to 78).

Totally inaccessible stopping times; classification (nos. 79 to 81). Quasi-left-continuous filtrations (nos. 82 to 83). Cross section theorems (nos 84 to 87). Sets with countable sections (no. 88).
Additional numbers on sets with countable sections (nos

Examples and supplements 143

Optional or predictable processes defined by limits (nos. 89 to 93). Canonical spaces (nos. 94 to 96), their predictable and optional σ-fields (nos. 97 to 98) and Galmarino's test (nos. 99 to 102). Decomposition of a stopping time (no. 103). Filtrations generated by a single stopping time (nos. 104 to 108).

APPENDIX TO CHAPTER III 156

Souslin schemes (nos. 75 to 77). Representation of Souslin and Lusin spaces as continuous images of Polish spaces (nos. 78 to 80). Isomorphisms between Lusin spaces (no. 80). Cross section theorem (nos. 81 to 82). The second separation theorem (83).

APPENDIX TO CHAPTER IV 163

Sets with un-countable sections (nos. 109 to 113). Derivatives of random sets (nos. 114 to 116). Sets with countable sections (nos. 117 to 118).

COMMENTS 169

INDEX OF TERMINOLOGY 175

INDEX OF NOTATION 181

BIBLIOGRAPHY 185

Preface

The titles of most books are meant to provide some information about their contents. So it is only fair to warn the reader that this volume contains little enough probability and no potential theory whatsoever [(1)]. Most of the probability should appear in a subsequent volume (martingale theory), potential theory still later (resolvents and semi-groups). As for the "and" between these two words, it is pushed so far into the future that we scarcely dare think about it. The true contents of this volume are some brief recollections of measure theory and the vocabulary of probability, and two long chapters, the first one on analytic sets and capacities, and the second one on the foundations of stochastic processes.

Why then this title ? In the year 1966, the second author had already published a volume called Probability and Potentials, containing eleven chapters which covered a much wider domain. It only lacked the "and", that is, the connection between potential theory and the theory of Markov processes. This was meant for a second volume, whose partial outline appeared in 1967 as a set of lecture notes on Markov processes. And now, instead of completing this second part to crown the edifice, we return to the very foundations of it. This may look absurd, but there are several reasons for it. First of all, the need for a reference book on Markov precesses and potential theory which was felt in those times was relieved by the publication, in 1968, of the treatise of Blumenthal and Getoor. Next, our whole theory has been since 1966 in a process of very rapid evolution. To take a few examples, in probability theory the first edition of this book contained the definition of a well-measurable process, but that of a predictable process, which is now so basic for stochastic integration, was only implicit there. On the potential side, a modest notion called "pseudo-réduite" (p. 247 of the French edition) was

---

(1) So we are following in the footsteps of our Master N. Bourbaki, whose structures fondamentales de l'analyse contain no analysis at all.

introduced with the somewhat despising comment "we aren't sure that the following theorem can be of any use". From the work of Mokobodzki on resolvents, we have learnt since that (pseudo)réduites are the key to the deeper results of potential theory. Finally, concerning the "and" part, the announcement at the beginning of the section on Ray resolvents was "the following results will not be used in later chapters, and can be omitted", while they would now be considered as fundamental. Examples of this kind could be multiplied.

The conditions of our work have also changed considerably since 1966. At that time, whereas potential theory and the theory of Markov processes were respectable areas of mathematics, people interested in the relation between them would scarcely outnumber half a score in the world. This is no longer the case (may be some credit for it can be ascribed to the first edition of this book which, for all its imperfections, has contributed to popularizing a number of ideas). There are just two names on the cover, but this shouldn't hide the fact that the new points of view presented here, or to be presented in later volumes, came into being through innumerable exchanges. The reader will gain some idea of it by perusing the volumes of the Strasbourg probability seminar, published every year since 1967 - and this is but the tip of the iceberg.

Thus the rapid evolution of the whole theory has discouraged us from building on the old foundations, and the support of an active mathematical environment has been an incentive to undertake again the full work from the start. Our publisher also has been full of understanding in his acceptance of a publication "by instalments", more informal than usual.

From the history of our theory we have also learnt some lessons, which we have tried to put to use in this new edition. In particular, we have tried to free ourselves from the attitude of many textbooks, which deal with mathematical truths as with eternal objects offered to our contemplation, from a world of pure Ideas where inflation is something unheard of. Truths are truths, but their value doesn't come from being printed on fine paper. Many immutable truths of 1966 have lost all interest and are now dead, while small remarks of 1966 have grown up and now shed light on large parts of our field.

So we have tried to put as much life as we could into the work, making digressions, adding comments and leaving some room for technical tricks and "useless" remarks. We must confess that the material may be considered arid, and that boredom has overcome us at times (at which places, the reader will probably know by his own weariness), but not too often.

We have preserved the organization of the first edition : within each chapter, all statements (whether theorems, definition or remarks) to which it may be useful to refer are sequentially numbered. So Theorem II. 31 (II denotes the chapter) may be followed by remarks II. 32 a) and b), and Definition II. 33. This is convenient for the reader (so we believe), but the cost to the authors of modifying a chapter that is almost completed becomes enormous. So we beg from our readers some indulgence for irregularities : "bis" and missing numbers, or maybe trivial remarks glorified with a number to prevent a gap in the numbering. Indexes of notation and terminology at the end of the volume are organized according to this system, not to page numbers. The bibliography is classified by alphabetical order of authors, but in the list of each author's publications (numbered [1], [2] ...) the order is purely random, as fits a probability book.

We should have dedicated the book to our wives, for keeping the children quiet while Daddy was working (or pretending to), but we got from Frank Knight the (secret) information that 1976 was the year of Doob's 65 th birthday (1). Now Doob's ideas inspired a great deal of the work in our field and in particular pervade the whole of our chapter IV. So it was only justice to write here:

DEDICATED TO J.L. DOOB ON HIS 65 TH BIRTHDAY

---

(1) Our hearty thanks go to Professor T.G. Kurtz, who helped us to prepare the final manuscript of the English edition. His comments, on mathematics and language, led to the elimination of many errors and obscurities.

CHAPTER 0

Notation

Notation from set theory 1

The complement of A is denoted by $\complement A$ or more often A^c. The notation $A \setminus B$ means $A \cap B^c$; $A \triangle B$ is the symmetric difference $(A\setminus B) \cup (B\setminus A)$. The set of all $x \in E$ with some property P is denoted by $\{x \in E : P(x)\}$ or, if there is no ambiguity, $\{x : P(x)\}$ or simply $\{P\}$.

The restriction of a function f to a set A is denoted by $f|_A$. Similarly, if $\mathcal{E}$ is a family of subsets, $\mathcal{E}|_A$ is the set of traces on A of elements of $\mathcal{E}$: explicitly $\mathcal{E}|_A = \{B \cap A,\ B \in \mathcal{E}\}$.

Closure of sets of subsets 2

We sometimes use sentences of the following form : the family $\mathcal{E}$ is closed under (...), where the brackets contain set-theoretic operation symbols, sometimes followed by the letters f, c, a, m, which abbreviate respectively : finite, countable, arbitrary, monotone. Two examples will suffice to clarify their meaning : "$\mathcal{E}$ <u>is closed under</u> $(\cup f, \cap a)$" means that finite unions (*) of elements of $\mathcal{E}$ and arbitrary intersections of elements of $\mathcal{E}$ still belong to $\mathcal{E}$; "$\mathcal{E}$ is <u>closed under</u> $(\cup mc,\ ^c)$" means that monotone countable unions of elements of $\mathcal{E}$ (i.e. unions of increasing sequences in $\mathcal{E}$) still belong to $\mathcal{E}$ and that complements of elements of $\mathcal{E}$ still belong to $\mathcal{E}$. Sets of subsets or functions are generally denoted by capital script letters.

The closure of a family of subsets $\mathcal{E}$ under $(\cup c)$ (resp. $(\cap c)$) is denoted by $\mathcal{E}_\sigma$ (resp. $\mathcal{E}_\delta$) - this notation is classical to set theory. We write $((\mathcal{E})_\sigma)_\delta = \mathcal{E}_{\sigma\delta}$.

Lattice notation 3

Let f and g be two real-valued functions. We write $f \vee g$ and $f \wedge g$ for sup(f,g) and inf(f,g). The notation f^+ and f^- has its classical meaning : $f^+ = f \vee 0$, $f^- = (-f) \vee 0$.

More generally, $\vee, \wedge$ denote least upper and greatest lower bounds : for example, the σ-field generated by the union of a family of σ-fields $\mathcal{F}_i$ is denoted by $\bigvee_i \mathcal{F}_i$.

(*) Bourbaki includes under finite unions the "empty union" and similarly for intersections. We do not use this convention.

4 Limits along $\mathbb{R}$ and $\mathbb{N}$

The notation $s \uparrow t$ means $s \to t$, $s \leq t$; $s \uparrow\uparrow t$ means $s \to t$, $s < t$; $s_n \uparrow t$, $s_n \uparrow\uparrow t$ is used similarly for sequences (s_n), with the additional meaning that (s_n) is increasing. Obvious changes are required if $\downarrow$ appears instead of $\uparrow$. The usual notations $\lim_{n\to\infty}$, $\liminf_{n\to\infty}$ for limits along $\mathbb{N}$ will be written simply as $\lim_n$, $\liminf_n$.

5 Integration theory

The word measure without further qualification always means "positive (1) countably additive set function on an abstract measurable space". We do not adhere to the convention that all measures considered are σ-finite: this would cost us too much generality in potential theory but we'll consider only countable sums of bounded measures (no difficulties relating to Fubini's theorem, for example, can arise with, such measures).

The notation $\lambda \vee \mu = \sup(\lambda,\mu)$, $\lambda \wedge \mu = \inf(\lambda,\mu)$, μ^+, μ^-, $|\mu| = \mu^+ + \mu^-$ has its classical meaning. $\|\mu\|$ denotes the total mass $<|\mu|,1>$ of μ (sometimes infinite). The integral of a function f with respect to a measure μ is denoted by $\int f(x)\mu(dx)$ (2) : often abridged into $\int f\mu$; $\frac{\mu}{\nu}$ denotes a Radon-Nikodym density, without "d". However when a measure μ on $\mathbb{R}$ appears as the derivative of an increasing function F, we use the standard notation with "d" for Stieltjes integrals $\int f(x)dF(x)$, and in particular $\int f(x)dx$ if $F(x) = x$.

If μ is a probability law, we often write $\mathbb{E}[f]$ for $\int f\mu$ and $\mathbb{E}[f,A]$ for $\int_A f\mu$ (especially when A is a complicated event).

6 Function spaces

If E is a topological space, $\mathcal{C}(E)$, $\mathcal{C}_b(E)$, $\mathcal{C}_c(E)$, $\mathcal{C}_0(E)$ denote the spaces of real-valued functions which are respectively continuous, bounded and continuous, continuous with compact support, continuous and tending to 0 at infinity (the latter when E is locally compact). Adjoining a $^+$ to this notation ($\mathcal{C}^+(E)$ and so on) enables us to denote the corresponding cones of positive functions. As usual $\mathcal{C}_c^\infty(E)$ denotes the space of infinitely differentiable functions with compact support (E then is assumed to be a manifeld, usually R^n).

If $(E,\mathcal{E})$ is a measurable space, the notation $m(\mathcal{E})$ (resp. $b(\mathcal{E})$) denotes the space of $\mathcal{E}$-measurable (resp. bounded $\mathcal{E}$-measurable) real functions. Spaces of measures are used only on a Hausdorff topological space E : $\mathcal{M}_b^+(E)$, $\mathcal{M}^+(E)$ then are the cones of bounded (resp. arbitrary) Radon measures on E and $\mathcal{M}_b(E)$, $\mathcal{M}(E)$ are the vector spaces generated by these cones.

---

(1) Positive means ≥ 0, according to European use.

(2) Also $\mu(f)$, $<\mu,f>$.

Topological spaces 7

A topological space E is *separable* if it contains a countable dense set. It is called *Polish* if there exists a distance compatible with its topology under which E is complete and separable. Locally compact spaces with countable base are called *LCC spaces*.

We write *l.s.c.* (*u.s.c.*) to abbreviate the expression *lower* (*upper*) *semi-continuous*.

Ordinals and transfinite induction 8

We consider arguments by transfinite induction as extremely convenient and intuitive and we use them freely. The following section contains everything our reader may need to understand our proofs.

Let $(J,\leq)$ be some non-countable well-ordered [(1)] set with a last element ε. Such a set "exists" : well-order any un-countable set (axiom of choice) and adjoin ε to it if necessary. For all $\alpha \in J$, let C_α be the set of $\beta \in J$ such that $\beta < \alpha$ (i.e. $\beta \leq \alpha$, $\beta \neq \alpha$).

The set of α such that C_α is un-countable contains ε by hypothesis, it is therefore non-empty and hence has a smallest element j_0. Let $I = C_{j_0}$ with the order induced by J. The ordered set I has the two following properties :

(1) it is well-ordered and non-countable ;

(2) for all $i \in I$, C_i is countable.

It can be shown that these two properties characterize the ordered set I up to isomorphism, and in particular that I is isomorphic to the standard set of all countable ordinals, but we do not need to know that. Having chosen I once and for all, we adopt the following terminology :

- the elements of I *are called* (*countable*) *ordinals* ;
- given $\alpha \in I$, it is very easy to see that the set of all $\beta \in I$ such that $\alpha < \beta$ is non-empty (otherwise I would be countable!). We denote by $\alpha + 1$ its smallest element and we call it the *successor* of α and α the *predecessor* of $\alpha + 1$ (it is uniquely determined as the largest element of $\{\beta : \beta < \alpha + 1\}$;
- the smallest element of I is denoted by 0, we write quite naturally $0 + 1 = 1$, $1 + 1 = 2$, $2 + 1 = 3, \ldots$ The least upper bound of the "integers" thus constructed is often denoted by ω ;
- an ordinal of the form $\alpha + 1$ is often called *a first kind* ordinal and an ordinal $\neq 0$ without a predecessor a *second kind* ordinal (or a *limit ordinal*). We now group in a single statement some less obvious properties of I.

---

(1) Recall that a *totally* ordered set is well-ordered if every non-empty subset contains a smallest element.

LEMMA. a) *For every ordinal* $\alpha \in I$, $\alpha \neq 0$, *there exists a strictly increasing mapping* f *of the interval* $[0,\alpha]$ *of* I *into* $\mathbb{R}$ *such that* $f(0) = 0$, $f(\alpha) = 1$.

b) *Conversely, for every increasing mapping* f *of the whole of* I *into* $\mathbb{R}$ *or* $\overline{\mathbb{R}}$, *there exists an ordinal* γ *such that* $f(\beta) = f(\gamma)$ *for all* $\beta \geq \gamma$.

c) *For every limit ordinal* α, *there exists an increasing sequence of ordinals* $\alpha_n < \alpha$ *such that* $\alpha = \sup_n \alpha_n$.

Proof : a) Let A be the set of ordinals $\beta \neq 0$ such that no such mapping f of $[0,\beta]$ into $\mathbb{R}$ exists. If A is non-empty, it has a smallest element α. Clearly α cannot have a predecessor and α is not 0. Hence α is a limit ordinal. For every $\beta < \alpha$, let f_β be a strictly increasing mapping of $[0,\beta]$ into $\mathbb{R}$ such that $f_\beta(0) = 0$, $f_\beta(\beta) = 1$ and let g_β be the mapping of $[0,\alpha]$ into $\mathbb{R}$ equal to f_β on $[0,\beta]$ and to 1 on $]\beta,\alpha]$. Since there are countably many $\beta < \alpha$, there exist numbers ε_β, all strictly positive, such that $\sum_\beta \varepsilon_\beta = 1$. Then the function $\sum_\beta \varepsilon_\beta g_\beta$ satisfies the statement on $[0,\alpha]$, contradicting the definition of α. Hence A is empty.

b) Let $A = \sup_\beta f(\beta)$. For all n let α_n be such that $f(\alpha_n) > A - \frac{1}{n}$ ($f(\alpha_n) > n$ if $A = +\infty$) and let $\gamma = \sup_n \alpha_n$. Then $f(\beta) = f(\gamma) = A$ for all $\beta \geq \gamma$.

c) Finally, let α be a limit ordinal and let f be a strictly increasing mapping of $[0,\alpha]$ into a bounded interval of $\mathbb{R}$. Let $c = \sup_{\beta<\alpha} f(\beta)$. It suffices to take α_n to be the smallest ordinal β such that $f(\beta) > c - \frac{1}{n}$.

In naïve language, the "principle of transfinite induction" can be stated as follows : let $P(\alpha)$ be a "property of an ordinal α" (in other words, a subset A of I: $P(\alpha)$ is true for α if and only if $\alpha \in A$), such that

(1) if P is true for α, P is true for $\alpha + 1$,

(2) if β is a limit ordinal and P is true for every ordinal $\alpha < \beta$, P is true for β.

(3) P is true for 0.

Then P *is true for all* $\alpha \in I$. This is obvious : Let A be the set of all α for which $P(\alpha)$ is not true. If A is non-empty, A has a first element which has no predecessor (1), is not a limit ordinal (2) and is not 0 (3). This is absurd, hence A must be empty.

This "principle" applies of course to every well-ordered set and not only to I.

Transfinite induction can be used in a slightly different way, to *construct* a function f on the set of all countable ordinals. One then applies the above argument, $P(\alpha)$ being the property "there exists on the interval $[0,\alpha]$ one and only one function f such that...", and the conclusion being "there exists on the whole of I one and only one function f such that...". Usually the dots... represent induction rules, defining $f(\alpha)$ when $f(\beta)$ is known for all $\beta < \alpha$.

9 Finally, a word about the (restricted) "continuum hypothesis". It is known that the following axiom:

I has the power of the continuum

(in other words, that it is possible to "enumerate" by means of I the points of $\mathbb{R}$, the sequences of integers,...) is independent of the usual axioms of set theory. Until now, the adoption or rejection of this axiom has been simply a matter of taste, with no really useful result of analysis depending on it. We shall see below several very beautiful consequences of the continuum hypothesis (due to Mokobodzki), which lead us to adopt it in this book with the same standing as the axiom of choice. See however the comments.

CHAPTER I

Measurable spaces

1. σ-FIELDS AND RANDOM VARIABLES

DEFINITION. Let Ω be a set. A σ-field on Ω is a family of subsets of Ω which contains 1
the empty set and which is closed under the operations $(\cup c, \cap c, {}^c)$.

The word σ-algebra is a synonym frequently used for σ-field (1). The ordered pair $(\Omega,\mathfrak{F})$ consisting of a set Ω and a σ-field $\mathfrak{F}$ of subsets of Ω is called a measurable space ; the elements of $\mathfrak{F}$ are called measurable or $\mathfrak{F}$-measurable sets. In the language of probability theory, they are called events.

The set Ω is then called the "sure event" and the empty set the "impossible event"; the operation of taking complements is called passing to the "opposite event". We sometimes use phrases such as : "the event A occurs", "the events A and B occur simultaneously", "the events A and B are incompatible", to express the assertions which we would write respectively in set theoretic terms : "$\omega \in A$", "$\omega \in A \cap B$" and "$A \cap B = \emptyset$". The reader will no doubt very quickly become familiar with such statements.

To understand this language and use it fully as an aid to intuition, the points of Ω can be conceived of as the possible results of drawing lots. Every subset A of Ω is then associated with an "event" in the usual sense of this term, a physical phenomenon which occurs when the result ω "falls in A". Among all possible "events", the σ-field $\mathfrak{F}$ contains those which are sufficiently simple to have a probability attributed to them.

Definition of random variables

DEFINITION. Let $(\Omega,\mathfrak{F})$, $(E,\mathcal{E})$ be two measurable spaces. A mapping f of Ω into E is 2
called measurable if

$$f^{-1}(A) \in \mathfrak{F} \quad \text{for all } A \in \mathcal{E} .$$

---

(1) A (Boolean) algebra is a set of subsets of Ω which contains $\emptyset$ and is closed under $(\cup f, \cap f, {}^c)$.

In the language of probability theory, we also say that f is a random variable (r.v.).

If there is any danger of ambiguity, we say explicitly: a random variable on $(\Omega,\mathcal{F})$ with values in $(E,\mathcal{E})$.

3 EXAMPLES

(1) Let E be the set consisting of the two numbers 0, 1, with the σ-field $\mathcal{E}$ of all subsets of E. A subset A of Ω is an event if and only if its characteristic function χ_A (equal to 1 on A and to 0 on $\Omega\backslash A$) is a random variable. Probabilists prefer to call this function the indicator of A and to denote it by I_A. We shall use this terminology in what follows.

(2) Let $(\Omega,\mathcal{F})$ be a measurable space and $\mathcal{G}$ a sub-σ-field of $\mathcal{F}$. The identity mapping of $(\Omega,\mathcal{F})$ onto $(\Omega,\mathcal{G})$ then is a random variable.

4 THEOREM. Let $(\Omega,\mathcal{F})$, $(G,\mathcal{G})$ and $(E,\mathcal{E})$ be three measurable spaces and $u:\Omega\to G$, $v:G\to E$ two random variables. The composite mapping $v\circ u$ then is a random variable.

5 DEFINITION. a) Let Ω be a set and $\mathcal{A}$ a family of subsets of Ω. The σ-field generated by $\mathcal{A}$, denoted by $\sigma(\mathcal{A})$, is the smallest σ-field of subsets of Ω containing $\mathcal{A}$.

b) Let Ω be a set and $(f_i)_{i\in I}$ a family of mappings of Ω into measurable spaces $(E_i,\mathcal{E}_i)_{i\in I}$. The σ-field generated by the mappings f_i, denoted by $\sigma(f_i, i\in I)$, is the smallest σ-field of subsets of Ω with respect to which all the mappings f_i are measurable.

The existence of the σ-fields described in these two definitions is obvious : on just takes the intersection (in $\mathcal{P}(\Omega)$) of all σ-fields for which the sets or functions considered are measurable - there is at least one such σ-field, mamely the σ-field consisting of all subsets of Ω.

There is moreover a close relation between parts (a) and (b) of 5 : the σ-field generated by a set of subsets is also generated by the indicators of these subsets ; the σ-field generated by the mappings f_i is also generated by the family of subsets of the form $f_i^{-1}(A_i)$, where for all i, A_i belongs to $\mathcal{E}_i$.

6 REMARKS. a) Let $(E,\mathcal{E})$ be a measurable space and f a mapping of E into Ω; then f is measurable with respect to the σ-field $\sigma(\mathcal{A})$ (5(a)) if and only if $f^{-1}(A)\in\mathcal{E}$ for all $A\in\mathcal{A}$ (the family of subsets B of Ω such that $f^{-1}(B)\in\mathcal{E}$ is a σ-field which contains $\mathcal{A}$ and hence also $\sigma(\mathcal{A})$).

Similarly, f is measurable with respect to the σ-field $\sigma(f_i, i\in I)$ on Ω (5(b)), if and only if each one of the mappings $f_i\circ f$ is measurable.

b) The σ-field generated by a family of functions $(f_i)_{i\in I}$ is identical to the union (in $\mathcal{P}(\Omega)$) of all the σ-fields $\sigma(f_i, i\in J)$, with J running through the family of all countable subsets of I.

7 EXAMPLES OF σ-FIELDS

(a) Let E be a topological space. The Borel σ-field of E, denoted by $\mathcal{B}(E)$, is the σ-field generated by the open sets of E. If F is a subspace of E, the elements of $\mathcal{B}(F)$ are the traces on F of the elements of $\mathcal{B}(E)$. When E is a space with a countable base $\mathcal{H}$, every open set is a union of a sequence of elements of $\mathcal{H}$ and $\mathcal{B}(E)$ is therefore generated by $\mathcal{H}$. For example, the Borel σ-field of the real line is generated by the open intervals with rational endpoints. If E and F are two topological spaces, a Borel mapping of E into F is any mapping of E into F which is measurable from $(E,\mathcal{B}(E))$ to $(F,\mathcal{B}(F))$. Every continuous mapping is Borel.

When we consider a topological space E as a measurable space without specifying the σ-field, we always mean the σ-field $\mathcal{B}(E)$.

(b) The continuous real-valued functions on E generate on E a smaller σ-field than $\mathcal{B}(E)$, called the Baire σ-field of E. It coincides with $\mathcal{B}(E)$ when E is metrizable : for if d is a distance on E and F is a closed set, then $F = \{x: f(x) = 0\}$, where f is the continuous function $x \mapsto d(x,F)$.

(c) The family of all subsets of $\mathbb{R}$ which are measurable in the Lebesgue sense is a σ-field richer than $\mathcal{B}(\mathbb{R})$.

(d) The family of all subsets of $\mathbb{R}$ which are either countable or have a countable complement, is a σ-field (generated by the sets $\{x\}$, $x \in \mathbb{R}$). We shall see in Chapter III that this σ-field presents "pathological" characteristics (no. III, 26).

Product σ-fields

8 DEFINITION. Let $(E_i,\mathcal{E}_i)_{i\in I}$ be a family of measurable spaces; let E denote the product set $\prod_{i\in I} E_i$ and X_i $(i \in I)$ the coordinate mappings. The σ-field $\sigma(X_i, i \in I)$ is called the product σ-field of the σ-fields $\mathcal{E}_i$ and denoted by $\prod_{i\in I} \mathcal{E}_i$.

We denote the product of two σ-fields by $\mathcal{E}_1 \times \mathcal{E}_2$. Many authors write $\mathcal{E}_1 \otimes \mathcal{E}_2$, $\bigotimes_{i\in I} \mathcal{E}_i$.

REMARKS. a) The product σ-field is also generated by the subsets of E of the form $\prod_{i\in I} A_i$, where $A_i \in \mathcal{E}_i$ for all $i \in I$ and $A_i = E_i$ except for a finite number of indices.

b) Let f_i $(i \in I)$ be mappings of a set Ω into measurable spaces $(E_i,\mathcal{E}_i)$ and f be the mapping $(f_i)_{i\in I}$ of Ω into the product set $\prod_{i\in I} E_i = E$. We give this set the product σ-field : we then have $\sigma(f_i, i \in I) = \sigma(f)$. With the same notation, suppose that Ω has been given a σ-field $\mathcal{F}$ and that each of the mappings f_i of Ω into E_i is measurable; then the mapping f of Ω into E is measurable.

c) If (E_n) is a finite or infinite sequence of topological spaces with countable bases $\mathcal{L}_n$ (in particular, separable metrizable spaces) then $\mathcal{B}(\prod_n E_n) = \prod_n \mathcal{B}(E_n)$. We may indeed assume that $E_n \in \mathcal{L}_n$ for all n. Then the sets of the form $\prod_n U_n$, where $U_n \in \mathcal{L}_n$ for all n, $U_n = E_n$ except for a finite number of indices, form a countable base of open sets of the topology of $\prod_n E_n$ and hence generate $\mathcal{B}(\prod_n E_n)$. On the other hand, they generate $\prod_n \mathcal{B}(E_n)$.

9 Atoms ; separable σ-fields

Let $(\Omega,\mathcal{F})$ be a measurable space. The atoms of $\mathcal{F}$ are the equivalence classes in Ω for the relation

$$(9.1) \qquad I_A(\omega) = I_A(\omega') \text{ for all } A \in \mathcal{F}.$$

Every real-valued measurable mapping on Ω (or, more generally, with values in a separable metrizable space), being a limit of elementary functions (17), is constant on atoms. The measurable space $(\Omega,\mathcal{F})$ is called Hausdorff [(1)] if the atoms of $\mathcal{F}$ are the points of Ω. If $(\Omega,\mathcal{F})$ is not Hausdorff, we define the associated Hausdorff space $(\dot{\Omega},\dot{\mathcal{F}})$ as follows : $\dot{\Omega}$ is the quotient space of Ω by the relation (9.1) and $\dot{\mathcal{F}}$ is the σ-field consisting of the images in $\dot{\Omega}$ of the elements of $\mathcal{F}$ under the canonical mapping of Ω onto $\dot{\Omega}$.

10 The measurable space $(\Omega,\mathcal{F})$ is called separable (or $\mathcal{F}$ alone is said to be separable) if there exists a sequence [(2)] of elements of $\mathcal{F}$ which generates $\mathcal{F}$. If the σ-field $\mathcal{F}$ is separable, generated by a sequence (A_n), the atom of $\mathcal{F}$ which contains the point $\omega \in \Omega$ is the intersection of those A_n or A_n^c which contain ω : the atoms are therefore measurable. Note that the Hausdorff space associated with a separable space is also separable.

11 Two measurable spaces are said to be isomorphic if there exists a bijection between them, which is measurable and has a measurable inverse (such a bijection is a measurable isomorphism ; between topological spaces given their Borel fields, it is also called a Borel isomorphism). Clearly a measurable space isomorphic to a separable metrizable space (with its Borel σ-field) is a separable Hausdorff space. Conversely,

THEOREM. Let $(E,\mathcal{E})$ be a separable Hausdorff measurable space. Then $(E,\mathcal{E})$ is isomorphic to a (not necessarily Borel) subspace of $\mathbb{R}$ with its Borel σ-field. More precisely, if (A_n) is a sequence of subsets of E generating $\mathcal{E}$, the mapping [(3)] f defined by

$$f(x) = \sum_n 3^{-n} I_{A_n}(x) \text{ for all } x \in E$$

is a measurable isomorphism of $(E,\mathcal{E})$ onto $(f(E),\mathcal{B}(f(E)))$.

Proof : Clearly f is a measurable bijection of E onto f(E). To show that its inverse is measurable, it suffices to show that the σ-field generated by f, which is contained in $\mathcal{E}$, is equal to $\mathcal{E}$, or equivalently that it contains each of the A_n. This follows from the fact that A_n is the inverse image under f of the $y \in [0,2]$ such

(1) In French "espace séparé". It seems that no "official" English terminology exists.
(2) The closure under $(\cup f, \cap f, {}^c)$ of a countable family of subsets still is countable hence a separable σ-field is generated by a countable Boolean algebra.
(3) The mapping $\frac{2}{3}f$, which takes values in the classical Cantor set, is sometimes called the "Marczewski indicator" of the sequence (A_n).

that the n-th digit of the expansion of y in base 3. 3 is equal to 1.

Let us show that, under suitable separability hypotheses, some useful sets are measurable.

THEOREM. Let $(\Omega,\mathcal{F})$ be a measurable space and $(E,\mathcal{E})$ be a separable Hausdorff measurable space. 12

(a) The diagonal of $E \times E$ belongs to the product σ-field $\mathcal{E} \times \mathcal{E}$.

(b) If f is a measurable mapping of Ω into E, the graph of f in $\Omega \times E$ belongs to the product σ-field $\mathcal{F} \times \mathcal{E}$.

(c) If f and g are measurable mappings of Ω into E, the set $\{f = g\}$ belongs to $\mathcal{F}$.

Proof : Let $\mathcal{A} = (A_n)$ be a countable Boolean algebra generating $\mathcal{E}$ and let D be the set of all $(m,n) \in \mathbb{N} \times \mathbb{N}$ such that $A_m \cap A_n = \emptyset$. Since $\mathcal{E}$ is Hausdorff, the diagonal of $E \times E$ is the complement of the union of the $A_m \times A_n$, where (m,n) runs through D, and hence belongs to $\mathcal{E} \times \mathcal{E}$. Assertion (c) follows from the fact that the set $\{f = g\}$ is the inverse image of the diagonal of $E \times E$ under the measurable mapping $\omega \to (f(\omega), g(\omega))$ of $(\Omega,\mathcal{F})$ into $(E \times E, \mathcal{E} \times \mathcal{E})$. Assertion (b) follows from (c) applied to $(\Omega',\mathcal{F}') = (\Omega \times E, \mathcal{F} \times \mathcal{E})$ and to the measurable mappings of Ω' into E, $f': (\omega,x) \mapsto f(\omega)$ and $g': (\omega,x) \mapsto x$; the graph of f is equal to the set $\{f' = g'\}$.

In fact, this theorem reduced by 11 to the classical special case where E is a separable metrizable space and $\mathcal{E} = \mathcal{B}(E)$. We shall adopt the more intuitive topological terminology, whenever possible, and say for instance, "let E be a separable metrizable space..." rather than "let $(E,\mathcal{E})$ be a separable Hausdorff measurable space..."

2. REAL-VALUED RANDOM VARIABLES

As usual, functions taking their values in $\overline{\mathbb{R}} = \mathbb{R} \cup \{+\infty,-\infty\}$ are called extended real-valued functions, while real-valued functions aren't allowed the values $\pm\infty$. 13

The word elementary random variable is used currently to describe a random variable taking either countably many values or more precisely finitely many values. The meaning below is the former, unless explicity stated.

First properties

The functions f, g,... below are assumed to be defined on the same measurable space $(\Omega,\mathcal{F})$.

Let f and g be two extended real-valued random variables. Then the functions $f \wedge g$, $f \vee g$, $f + g$, fg (if everywhere defined) are random variables (1). 14

Let $(f_n)_{n \in \mathbb{N}}$ be a pointwise convergent sequence of extended real-valued random 15

(1) This is a chance to recall that the convention $0.\infty = 0/0 = 0$ is universally adopted in integration theory.

variables and let $f = \lim_n f_n$. Then the function f is a random variable.

This property extends to random variables with values in a metric space E : if d is the distance and F is closed in E, the set $\{f \in F\}$ is equal to $\{x : d(f_n(x),F) \underset{n \to \infty}{\to} 0\}$.

16 Let $(f_n)_{n \in \mathbb{N}}$ be a sequence of extended real-valued random variables. The convergence set of the sequence (f_n) is measurable.

This property extends to random variables with values in a subspace E of a complete metric space F if E is Borel in F : the set A of $\omega \in \Omega$ such that $(f_n(\omega))$ is a Cauchy sequence in F belongs to $\mathcal{F}$. The mapping f of Ω into F defined by $f(\omega) = \lim_n f_n(\omega)$ if $\omega \in A$ and $f(\omega) = x_0$ if $\omega \notin A$, where x_0 is any point of $F\setminus E$, is measurable ; the convergence set of (f_n) in E then is equal to $A \cap f^{-1}(E)$.

17 An extended real-valued function f is measurable, if and only if there exists a sequence (f_n) of measurable elementary functions which increases to f.

The following explicit sequence is often useful (1). It is known as "Lebesgue's approximation of f"

$$f_n = \sum_{k \in \mathbb{Z}} k2^{-n} I_{\{k2^{-n} < f \le (k+1)2^{-n}\}} + (-\infty) I_{\{f=-\infty\}}$$

Observe that it converges uniformly to f. If we replace the word "increases" by "converges uniformly", 17 extends to random variables with values in a separable metric space E, given its Borel σ-field (here generated by the open balls). For let $(x_n)_{n \in N}$ be a sequence dense in E and let B_n be the open ball of centre x_n and radius ε. Set $C_n = B_n \setminus (\bigcup_{p<n} B_p)$; the sets C_n then are Borel and disjoint and cover E. If g denotes the elementary function equal to x_n on $f^{-1}(C_n)$, g is measurable and its distance to f is less than ε.

18 The following theorem is due to Doob ([1], p. 603). It shows that the notion of a $\sigma(f)$-measurable random variable may be replaced by the less abstract notion of a measurable function of f. By 16 and 17, this extends easily to functions g taking values in a complete separable measurable space : details are left to the reader.

THEOREM. Let f be a random variable defined on $(\Omega, \mathcal{F})$ with values in a measurable space $(E, \mathcal{E})$ and g be a real-valued function defined on Ω. Then g is $\sigma(f)$-measurable, if and only if there exists on E a real-valued random variable h such that $g = h \circ f$.

Proof : The condition is obviously sufficient. To show that it is also necessary, we begin with the case where g assumes only countably many values a_n $(n \in \mathbb{N})$. Since the sets $A_n = \{g = a_n\}$ are $\sigma(f)$-measurable, they are of the form $f^{-1}(B_n)$ with $B_n \in \mathcal{E}$. Let $C_n = B_n \setminus (\bigcup_{p<n} B_p)$: these sets belong to $\mathcal{E}$ and are disjoint. On the other hand

$$f^{-1}(C_n) = A_n \setminus (\bigcup_{p<n} A_p) = A_n.$$

(1) If f is positive, $f_n = 2^{-n} \sum_{k>0} I_{\{f > k2^{-n}\}}$

Then let h be the function on E which takes the values a_n on C_n and 0 (for example) on $E \setminus (\bigcup_n C_n)$. It is obvious that $h \circ f = g$.

We now pass to the general case. By 17, there exists a sequence (g_n) of elementary $\sigma(f)$ measurable random variables converging to g, and g_n is of the form $h_n \circ f$ by the above. Let H be the convergence set of the sequence (h_n) : H is $\mathcal{E}$-measurable and contains $f(\Omega)$. We set

$$h(\omega) = \lim_n h_n(\omega) \quad \text{for } \omega \in H$$

$$h(\omega) = 0 \qquad\quad \text{for } \omega \in H.$$

The function thus constructed is the required function.

The "monotone class theorem"

The two theorems 19 and 21 are extremely useful.

THEOREM. *Let $\mathcal{C}$ be a family of subsets of Ω containing $\emptyset$ and closed under* $(\cup f, \cap f)$. *Let $\mathcal{M}$ be a family of subsets of Ω containing $\mathcal{C}$, and closed under* $(\cup mc, \cap mc)$ ($\mathcal{M}$ *is a "monotone class"*). *Then $\mathcal{M}$ contains the closure $\mathcal{S}$ of $\mathcal{C}$ under* $(\cup c, \cap c)$. (19)

If in addition $\mathcal{C}$ is closed under the operation $\complement$, *$\mathcal{M}$ contains the σ-field generated by $\mathcal{C}$.*

Proof :[1] To abbreviate, we call any set of subsets closed under $(\cup f, \cap f)$ a *horde*. Let $\mathcal{H}$ be a maximal horde among the hordes contained in $\mathcal{M}$ and containing $\mathcal{C}$ (Zorn's Lemma). We show that $\mathcal{H}$ is closed under $(\cup c, \cap c)$. Let (A_n) be a decreasing sequence[2] of elements of $\mathcal{H}$ and let $A = \bigcap A_n$; the family of all subsets of the form $(H \cap A) \cup H'$, with $H \in \mathcal{H} \cup \{\Omega\}$ and $H' \in \mathcal{H}$, is a horde containing $\mathcal{H}$ (take $H = \emptyset$) and A (take $H = \Omega$, $H' = \emptyset$) and contained in $\mathcal{M}$. Since $\mathcal{H}$ is maximal, this horde is identical to $\mathcal{H}$ and hence $A \in \mathcal{H}$. In other words $\mathcal{H}$ is closed under $(\cap c)$, so that $\mathcal{H}$ contains the closure of $\mathcal{C}$ under $(\cap c)$. The argument is similar for $(\cup c)$.

Let $\mathcal{J}$ be the set of all $A \in \mathcal{S}$ such that $A^c \in \mathcal{S}$; if the complement of every element of $\mathcal{C}$ belongs to $\mathcal{C}$ (or more generally to $\mathcal{S}$), $\mathcal{J}$ contains $\mathcal{C}$ and, in particular, $\emptyset \in \mathcal{J}$. Obviously $\mathcal{J}$ is a σ-field contained in $\mathcal{M}$, and the last sentence in the statement of the theorem follows.

The following example (which encroaches slightly on the beginning of chapter II) illustrates Theorem 19.

THEOREM. *Let $\mathcal{F}_0$ be a set of susbets of Ω, closed under the operations* $(\cup f, {}^c)$ [3]; *Let $\mathbb{P}$ and $\mathbb{P}'$ be two probability laws on* $\mathcal{F} = \sigma(\mathcal{F}_0)$ such that $\mathbb{P}(A) = \mathbb{P}(A)$ for all $A \in \mathcal{F}_0$. Then $\mathbb{P}$ and $\mathbb{P}'$ are equal on $\mathcal{F}$. 20

---

(1) For proofs not using Zorn's Lemma, see Chung [1], p. 17.

(2) We need only consider such a sequence since $\mathcal{H}$ is closed under $(\cap f)$

(3) That is, a Boolean algebra (no. 1).

Proof : Apply 19 with $\mathcal{C} = \mathcal{F}_0$ and with $\mathcal{M}$ denoting the set of elements A of $\mathcal{F}$ such that $\mathbb{P}(A) = \mathbb{P}'(A)$.

Here is the functional form of the monotone class theorem. We first give the statement we use most often, and then some variants of it.

21 THEOREM. Let $\mathcal{H}$ be a vector space of bounded real-valued functions defined on Ω, which contains the constants, is closed under uniform convergence and has the following property : for every uniformly bounded increasing sequence of positive functions $f_n \in \mathcal{H}$, the function $f = \lim_n f_n$ belong to $\mathcal{H}$.

Let $\mathcal{C}$ be a subset of $\mathcal{H}$ which is closed under multiplication. The space $\mathcal{H}$ then contains all bounded functions measurable with respect to the σ-field $\sigma(\mathcal{C})$.

Proof : Let $\mathcal{C}'$ be the algebra generated by the function 1 and the elements of $\mathcal{C}$; clearly $\mathcal{C}' \subset \mathcal{H}$. Zorn's Lemma allows us to choose a maximal element $\mathcal{A}_0$ of the set of algebras $\mathcal{A}$ satisfying the inclusion $\mathcal{C}' \subset \mathcal{A} \subset \mathcal{H}$. It is known (1) that the function $x \mapsto |x|$ can be uniformly approximated by polynomials on every compact interval of $\mathbb{R}$; the algebra $\mathcal{A}_0$ is obviously closed under uniform convergence and contains the constants, hence it is closed under the operation $f \mapsto |f|$, and therefore also under the operations $\vee$ and $\wedge$. Let g be the limit of a uniformly bounded increasing sequence of positive elements of $\mathcal{A}_0$; it is easily verified that the algebra generated by $\mathcal{A}_0$ and g is contained in $\mathcal{H}$: hence it is identical to $\mathcal{A}_0$ and $g \in \mathcal{A}_0$.

Let $\mathcal{S}$ be the family of all subsets of Ω whose indicators belong to $\mathcal{A}_0$; since $\mathcal{A}_0$ is an algebra, $\mathcal{S}$ is closed under the operations $(\cap f, {}^c)$. The closure of $\mathcal{A}_0$ under monotone convergence implies that $\mathcal{S}$ is a σ-field and, by virtue of 17, that $\mathcal{A}_0$ contains all the $\mathcal{S}$-measurable bounded functions. It remains to show that $\mathcal{S}$ contains $\sigma(\mathcal{C})$; it obviously suffices to show that $\mathcal{S}$ contains, for every function $f \in \mathcal{A}_0$, the set $B = \{\omega : f(\omega) \geq 1\}$. But the function $g = (f \wedge 1)^+$ belongs to $\mathcal{A}_0$ and the indicator of B is the limit of the decreasing sequence of functions g^n. This concludes the proof.

22 VARIANTS OF THEOREM 21. The "monotone class theorem" is one of the basic results of probability theory. To understand the intersest of its variants, let us assume we want to show that some property $\mathbb{P}$ is true of all bounded functions measurable with respect to some σ-field $\mathcal{J}$. We know how to prove $\mathbb{P}$ for some class $\mathcal{C}$ of functions which generates $\mathcal{J}$. We also know how $\mathbb{P}$ behaves under monotone convergence. Then letting $\mathcal{H}$ denote the set of bounded $\mathcal{J}$-measurable functions which satisfy $\mathbb{P}$, we expect that some variant of the monotone class theorem will tell us that $\mathcal{H}$ contains all the $\mathcal{J}$-measurable bounded functions.

---

(1) The Taylor series of $(1-z)^{1/2}$ converges uniformly on $[-1,+1]$. The function $|x| = (1-(1-x^2))^{1/2}$ can therefore be uniformly approximated by polynomials on $[-1,+1]$ (write $z = 1-x^2$).

The statement of 21 is appropriate to *linear* properties : if $\mathcal{P}$ is true for two functions f, g, it is still true for af + bg. In this case, little is assumed about $\mathcal{C}$. We repeat the hypotheses :

(22.1) (a) $\mathcal{H}$ is a *vector space* of bounded functions, closed under bounded monotone convergence and uniform convergence and containg 1 :

(b) $\mathcal{C}$ is closed under multiplication.

The same proof yields the same conclusion under the following hypotheses:

(22.2) (a) $\mathcal{H}$ is a *set* of bounded functions, closed under bounded monotone convergence and uniform convergence ;

(b) $\mathcal{C}$ is an *algebra* and $1 \in \mathcal{C}$.

(The condition $1 \in \mathcal{C}$ can in fact be weakened by assuming for example that there exist $f_n \in \mathcal{C}$ which increase to 1).

Uniform convergence just serves to pass from closure under multiplication to closure under the operations $\wedge, \vee$. Hence a variant without uniform convergence :

(22.3) (a) $\mathcal{H}$ is a set which is closed under bounded monotone convergence ;

(b) $\mathcal{C}$ is a $\wedge$-closed vector space and $1 \in \mathcal{C}$.

Here we must change the proof a little, replacing algebras $\mathcal{A}$ cointaining $\mathcal{C}$ by $\wedge$-closed vector spaces containing $\mathcal{C}$. If $\mathcal{A}_0$ is such a maximal space in $\mathcal{H}$, $\mathcal{A}_0$ is closed under monotone convergence. To conclude, it is necessary to know that if $g \in \mathcal{A}_0$ is positive, then $g^n \in \mathcal{A}_0$: this is very easy, since the convex function $x \mapsto x^n$ is the upper envelope of a sequence of affine functions.

As for 19, we now illustrate Theorem 21 by applications which anticipate Chapter II.

THEOREM. *Let* E *be a metric space given its Borel* σ-*field*. 23

(a) *Let* $\mathbb{P}$ *and* $\mathbb{P}'$ *be two probability laws such that* $\int f\mathbb{P} = \int f\mathbb{P}'$ *for every bounded continuous function* f. *Then* $\mathbb{P} = \mathbb{P}'$.

(b) *For every bounded Borel function* f *and all* $\varepsilon > 0$, *there exist two bounded functions* f' *and* f" *which are respectively u.s.c. and l.s.c., such that* $f' \le f \le f''$ *and* $\int (f'' - f')\mathbb{P} < \varepsilon$.

Proof : For (a), apply 21 with $\mathcal{C}$ the algebra of bounded continuous functions and $\mathcal{H}$ the set of bounded Borel functions f such that $\int f\mathbb{P} = \int f\mathbb{P}'$. We know that $\mathcal{C}_b(E)$ generates $\mathcal{B}(E)$ (15).

Here the convenient property certainly is closure under multiplication, as is shown by the special case where $E = \mathbb{R}^n$ and $\mathcal{C}_b(E)$ is replaced by the set of *bounded infinitely differentiable* functions. We leave it to the reader to prove the same result for *infinitely differentiable* functions *with compact support* : in the case of two probability laws $\mathbb{P}$ and $\mathbb{P}'$, or more generally of two locally bounded measures μ and μ' on $\mathbb{R}^n$.

For (b), we take for $\mathcal{C}$ the set of all bounded continuous functions and for $\mathcal{H}$

the set of all bounded Borel functions possessing the above stated approximation property. We then apply the form (22.3) of the theorem, which avoids uniform convergence. To show that $\mathcal{H}$ is closed under bounded monotone convergence, we consider an increasing sequence f_n of elements of $\mathcal{H}$ which are uniformly bounded and the corresponding u.s.c. functions f'_n and l.s.c. functions f''_n such that $f'_n \leq f_n \leq f''_n$ and $\int (f''_n - f'_n)\mathbb{P} < \varepsilon 2^{-n-2}$. We write $f = \lim_n f_n$, $f'' = \sup_n f''_n$, $f'_1 = \sup_n f'_n$ and verify that $f'_1 \leq f \leq f''$, $\int (f'' - f'_1)\mathbb{P} < \varepsilon/2$. The function f'' is l.s.c. but the function f'_1 is not u.s.c. : it is necessary to take f' to be a function $\sup_{n \leq N} f'_n$, where N is chosen sufficiently large so that $\int (f'_1 - f')\mathbb{P} < \varepsilon/2$.

Here is another example of the use of Theorem 21, useful in the theory of Markov processes.

24 THEOREM. Let $(\Omega,\mathcal{F},\mathbb{P})$ be a probability space and X and Y two random variables with values in a separable metric space E. To check that X = Y $\mathbb{P}$-a.s., it suffices to check that, for every pair (f,g) of bounded continuous functions on E,

(24.1) $$\mathbb{E}[f(X)g(Y)] = \mathbb{E}[f(X)g(X)].$$

Proof : Let $\mathcal{H}$ be the set of all bounded Borel functions h(x,y) on E × E such that $\mathbb{E}[h(X,X)] = \mathbb{E}[h(X,Y)]$: $\mathcal{H}$ is a vector space closed under bounded monotone convergence and uniform convergence. Let $\mathcal{C}$ be the set (closed under multiplication) of all functions of the form $(x,y) \mapsto f(x)g(y)$, where f and g are continuous and bounded on E (1). Formula (24.1) tells that $\mathcal{C} \subset \mathcal{H}$ and we know that $\mathcal{C}$ generates the σ-field $\mathcal{B}(E) \times \mathcal{B}(E) = \mathcal{B}(E \times E)$. By 21, $\mathcal{H}$ contains all bounded Borel functions. We conclude by taking h(x,y) to be the indicator of the complement of the diagonal.

(1) The function $(x,y) \mapsto f(x)g(y)$ is frequently denoted by $f \otimes g$.

CHAPTER II

Probability laws and mathematical expectations

As said in the introduction, we assume that our reader is familiar with the more classical parts of measure theory. The first part of this chapter is therefore simply a summary, intended to present the terminology of probability theory. We resume giving complete proofs in the paragraph devoted to uniform integrability.

1. A SUMMARY OF INTEGRATION THEORY

DEFINITION. *A* probability law *on a measurable space* $(\Omega,\mathfrak{F})$ *is a measure* $\mathbb{P}$ *defined on* 1
$\mathfrak{F}$, *which is positive and has a total mass of* 1.

The triple $(\Omega,\mathfrak{F},\mathbb{P})$ is called a *probability space*.

In other words, $\mathbb{P}$ is a positive function defined on $\mathfrak{F}$ such that $\mathbb{P}(\Omega) = 1$, which 2
satisfies the following property ("countable additivity") : $\mathbb{P}(\bigcup_n A_n) = \sum_n \mathbb{P}(A_n)$ for every sequence $(A_n)_{n \in \mathbb{N}}$ of disjoint events.

The number $\mathbb{P}(A)$ is called the *probability* of the event A. An event whose probability is equal to 1 is said to be *almost sure*. Let f and g be two random variables defined on $(\Omega,\mathfrak{F})$ with values in the same measurable space $(E,\mathcal{E})$. If the set $\{\omega : f(\omega) = g(\omega)\}$ is an event (1) of probability 1, we write

$$f = g \quad \text{a.s.}$$

where "a.s." is an abbreviation of "almost surely". Similarly, we shall write "A = B a.s." to express that two events A and B differ only by a set of zero probability. More generally, we use the expression "almost surely" in the same way as people use "almost everywhere" in measure theory. In fact probabilists freely use the vocabulary of measure theory alongside their own : this enables them to avoid repetition and makes their books very pleasant to read.

A probability space $(\Omega,\mathfrak{F},\mathbb{P})$ is called *complete* if every subset A of Ω which is 3
contained in a $\mathbb{P}$-negligible set belongs to the σ-field $\mathfrak{F}$ (and then necessarily $\mathbb{P}(A) = 0$). We shall return to this notion in 32 and prove there that any probability space can be completed.

EXAMPLES. (a) Let I be the interval $[0,1]$. Let us set, for every $A \in \mathcal{B}(I)$: 4

$$\mathbb{P}(A) = \int_A dx \quad \text{(Lebesgue measure).}$$

---

(1) This is the case if $(E,\mathcal{E})$ is separable and Hausdorff (I.12).

Then $\mathbb{P}$ is a probability law on I. $\mathbb{P}$ is not complete ; it becomes so when extended to the σ-field of Lebesgue measurable sets.

b) Let $(\sigma,\mathcal{F})$ be a measurable space and x be a point of Ω. We denote by ε_x the probability law defined by :

$$\varepsilon_x(A) = I_A(x) \quad (A \in \mathcal{F}).$$

This law is also called the degenerate law at x or the unit mass at x. More generally a law $\mathbb{P}$ on a measurable space $(\Omega,\mathcal{F})$ is said to be degenerate if $\mathbb{P}(A) = 0$ or 1 for all $A \in \mathcal{F}$. Every real-valued random variable then is a.s. equal to a constant.

Mathematical expectations

5 DEFINITION. Let $(\Omega,\mathcal{F},\mathbb{P})$ be a probability space and f be an integrable real-valued random variable. The integral $\int_\Omega f(\omega)P(d\omega)$ is called the mathematical expectation of the random variable f and is denoted by the symbol $\mathbb{E}[f]$.

We shall henceforth omit the adjective "mathematical".

We give few details on integration theory proper. We just state the two theorems which are most often used and make a few remarks.

6 LEBESGUE'S THEOREM (the dominated convergence theorem). Let $(f_n)_{n \in \mathbb{N}}$ be a sequence or real-valued random variables which converges almost surely [1], and let f be a random variable a.s. equal to $\lim_n f_n$. If the f_n are bounded in absolute value by some integrable function, f is integrable and $\mathbb{E}[f] = \lim_n \mathbb{E}[f_n]$.

Given a positive random variable f, finite or not, which is not integrable, we use the convention $\mathbb{E}[f] = +\infty$. Then the following theorem holds.

7 FATOU'S LEMMA. Let $(f_n)_{n \in N}$ be a sequence of positive random variables ; then we have :

$$\mathbb{E}[\liminf_n f_n] \leq \liminf_n \mathbb{E}[f_n].$$

This inequality can be replaced by equality when the sequence is increasing, whether the integrals are finite or not. This last result is known as Lebesgue's monotone convergence theorem.

8 In conformity with Bourbaki's notation, we denote by $\mathcal{L}^p(\Omega,\mathcal{F},\mathbb{P})$ (or simply $\mathcal{L}^p$) the vector space of real-valued random variables whose p-th power is integrable $(1 \leq p < \infty)$ and by L^p the quotient space of $\mathcal{L}^p$ by the equivalence relation defined by almost sure equality. For every real-valued measurable function f, we set

$$\|f\|_p = (\mathbb{E}[|f|^p])^{\frac{1}{p}} \quad \text{(possibly } +\infty).$$

Similarly, we denote by $\mathcal{L}^\infty(\Omega,\mathcal{F})$ the space (independent of $\mathbb{P}$) of bounded random variables, with the norm of uniform convergence, and by $L^\infty(\Omega,\mathcal{F},\mathbb{P})$ the quotient space of $\mathcal{L}^\infty$ by the same equivalence relation. The norm of an element f of L^∞ (the essential supremum of $|f|$) is denoted by $\|f\|_\infty$.

---

(1) Or even only in probability (see 10).

We shall use without further reference the following properties of the spaces L^p : the fact that L^p is a Banach space (see for example Dunford-Schwartz [1], p. 146) ; Hölder's inequality (ibid. p. 119) ; the fact that the dual of L^1 is L^∞ (ibid. p. 289). Another necessary result is the Radon-Nikodym theorem(ibid. p. 176), which will also be established in Chapter V as an application of martingale theory.

The following two remarks are useful 9

(a) Let f be an integrable random variable which is measurable with respect to a sub-σ-field $\mathcal{G}$ of $\mathcal{F}$. Then f is a.s. positive, if and only if

$$\int_A f(\omega)\mathbb{P}(d\omega) \geq 0 \quad \text{for all} \quad A \in \mathcal{G} .$$

(Take A to be the event $\{f < 0\}$).

It follows in particular that two integrable random variables f and g which are both $\mathcal{G}$-measurable and have the same integral on every set of $\mathcal{G}$ are a.s. equal.

(b) Let f and g be two integrable random variables; we say that f and g are <u>orthogonal</u> if the product f.g is integrable and has zero expectation. Let $\mathcal{G}$ denote a sub-σ-field of $\mathcal{F}$, U be the closed subspace of L^1 consisting of all classes of $\mathcal{G}$-measurable random variables, and V be the subspace of L^∞ consisting of all classes of bounded random variables orthogonal to every element of U. It follows from the Hahn-Banach theorem that every random variable $f \in \mathcal{L}^1$ orthogonal to every element of V is a.s. equal to a $\mathcal{G}$-measurable function.

Convergence of random variables

We now recall, restricting ourselves to the case of sequences, the main types 10
convergence of real-valued random variables (1).

Let (f_n) be a sequence of random variables defined on $(\Omega,\mathcal{F},\mathbb{P})$. We say that the sequence (f_n) converges to a random variable f :

- <u>almost surely</u> if $\mathbb{P}\{\omega : f_n(\omega) \to f(\omega)\} = 1$,
- <u>in probability</u> if $\lim_n \mathbb{P}\{\omega : |f_n(\omega) - f(\omega)| > \varepsilon\} = 0$ for all $\varepsilon > 0$,
- <u>in the strong sense in</u> L^p if the f_n and f belong to $\mathcal{L}^p$ and $\lim_n \mathbb{E}[|f_n - f|^p] = 0$,
- <u>in the weak sense in</u> L^1 (or alternatively : in the sense of the topology $\sigma(L^1,L^\infty)$)
- if the f_n and f belong to $\mathcal{L}^1$ and, for every random variable $g \in \mathcal{L}^\infty$, $\lim_n \mathbb{E}[f_n.g] = \mathbb{E}[f.g]$,
- <u>in the weak sense in</u> L^2 (or alternatively ; in the sense of the topology $\sigma(L^2,L^2)$) if the f_n and f belong to $\mathcal{L}^2$ and, for every random variable $g \in \mathcal{L}^2$, $\lim_n \mathbb{E}[f_n.g] = \mathbb{E}[f.g]$.

We shall return to weak convergence in L^1 in the section concerning uniform integrability. We just recall here that almost sure convergence and strong convergence in L^p imply convergence in probability and that every sequence which converges

---

(1) Or a.s. finite extended real-valued. The definitions relating to convergence in probability need slight modification for r.v. which are not a.s. finite.

in probability, contains a subsequence which converges almost surely. More precisely, let us set fo every real-valued random variable f

$$\pi[f] = \mathbb{E}[|f| \wedge 1].$$

Then the function $(f,g) \mapsto \pi[f-g]$ is a pseudo-metric which defines convergence in probability ; if the sequence (f_n) satisfies the property

$$\sum_n \pi[f_n - f_{n+1}] < \infty$$

it converges in probability and almost surely (see for example : Dunford and Schwartz [1], p. 150).

Image laws

11 DEFINITION. Let $(\Omega,\mathcal{F},\mathbb{P})$ be a probability space, $(E,\mathcal{E})$ be a measurable space and f be a random variable from Ω to E. The image law of $\mathbb{P}$ under f, denoted by $f(\mathbb{P})$, is the law Q on $(E,\mathcal{E})$ defined by :

$$Q(A) = \mathbb{P}(f^{-1}(A)) \quad (A \in \mathcal{E}).$$

This law is also called the law of or the distribution of f.

Let g be a measurable mapping of $(E,\mathcal{E})$ into a measurable space $(G,\mathcal{G})$. We have the obvious equation :

$$g(f(\mathbb{P})) = g \circ f)(\mathbb{P})$$

("transitivity of image laws").

12 THEOREM.-Let h be a real-valued random variable on $(E,\mathcal{E})$; h is Q-integrable if and only if $h \circ f$ is $\mathbb{P}$-integrable and then :

$$\int_E h(x)Q(dx) = \int_\Omega h \circ f(\omega)\mathbb{P}(d\omega).$$

Integration of probability laws ; Fubini's Theorem

13 DEFINITION. Let $(\Omega,\mathcal{F})$ and $(E,\mathcal{E})$ be two measurable spaces. A family $(P_x)_{x \in E}$ of probability laws on $(\Omega,\mathcal{F})$ is said to be $\mathcal{E}$-measurable if the function $x \mapsto P_x(A)$ is $\mathcal{E}$-measurable for all $A \in \mathcal{F}$.

Given such a family $(\mathbb{P}_x)_{x \in E}$, we have the following statement :

14 FUBINI'S THEOREM. Let Q be a probability law on $(E,\mathcal{E})$. Let $(U,\mathcal{U})$ denote the measurable space $(E \times \Omega, \mathcal{E} \times \mathcal{F})$.

(1) Let f be a real-valued random variable defined on $(U,\mathcal{U})$. Each one of the partial mappings $x \mapsto f(x,\omega)$, $\omega \mapsto f(x,\omega)$ is measurable on the corresponding factor space.

(2) There exists one and only one probability law $\mathbb{S}$ on $(U,\mathcal{U})$ such that, for all $A \in \mathcal{E}$ and $B \in \mathcal{E}$,

$$\mathbb{S}(A \times B) = \int_A \mathbb{P}_x(B)Q(dx). \tag{14.1}$$

(3) Let f be a positive [1] random variable on $(U,\mathcal{U})$. The function

$$x \mapsto \int_\Omega f(x,\omega)\mathbb{P}_x(d\omega).$$

(1) Recall that the integral has been defined for all positive measurable function (cf. 6).

is $\mathcal{E}$-measurable and :

(14.2) $$\int_U f(x,\omega)\mathbb{S}(dx,d\omega) = \int_E Q(dx)\int_\Omega f(x,\omega)\mathbb{P}(d\omega).$$

This relation still holds true if f is $\mathbb{S}$-integrable ; but one can then only assert that $\omega \mapsto f(x,\omega)$ is $\mathbb{P}_x$-integrable for Q-almost all $x \in E$.

REMARKS. (a) If f is neither positive nor $\mathbb{S}$-integrable, the right-hand side of (14.2) may be meaningful without the left-hand side being so.

(b) If all the $\mathbb{P}_x$ are equal to the same law $\mathbb{P}$, the law $\mathbb{S}$ is called the product (law) of $\mathbb{Q}$ and $\mathbb{P}$ and denoted by $\mathbb{Q} \otimes \mathbb{P}$. The probability space $(U,\mathcal{U},\mathbb{Q} \otimes \mathbb{P})$ is not complete in general. Fubini's Theorem is often stated for product laws only and in a slightly different form : assume that the factor spaces are complete and that f is measurable on the completed product space ; assertion (1) then is no longer true, but still the partial mappings $x \mapsto f(x,\omega)$ (resp. $\omega \mapsto f(x,\omega)$) are $\mathcal{E}$-measurable (resp. $\mathcal{F}$-measurable) for $\mathbb{Q}$-almost all $x \in E$ (resp. for $\mathbb{P}$-almost all $\omega \in \Omega$).

(c) The définition of the product of finitely many probability laws is obvious. We do not study here infinite products, which, however, are examples of inverse limits of probability laws, see Chapter III.

DEFINITION. In the notation of 14, the integral of the family $\mathbb{P}_x$ with respect to Q, 15
denoted by $\int_E \mathbb{P}_x Q(dx)$, is the image law of $\mathbb{S}$ under the projection mapping of $E \times \Omega$ onto Ω.

By combining 12 and 14 we get the following theorem.

THEOREM. Let $\mathbb{P}$ denote the law $\int_E \mathbb{P}_x \mathbb{Q}(dx)$ and f be a positive random variable on $(\Omega,\mathcal{F})$. 16
Then the function $x \mapsto \int_\Omega f(\omega)\mathbb{P}_x(d\omega)$ is $\mathcal{E}$-measurable and

$$\int_\Omega f(\omega)\mathbb{P}(d\omega) = \int_E Q(dx)\int_\Omega f(\omega)\mathbb{P}_x(d\omega).$$

This relation is also true for every $\mathbb{P}$-integrable random variable f ; However f is $\mathbb{P}_x$-integrable only for Q-almost all $x \in E$, so that $\int_\Omega f(\omega)\mathbb{P}_x(d\omega)$ is defined $\mathbb{Q}$-a.s., and no longer on the whole of E.

2. SUPPLEMENT ON INTEGRATION

Uniformly integrable random variables

All the random variables considered in this section are real-valued and defined on the same probability space $(\Omega,\mathcal{F},\mathbb{P})$ (1).

DEFINITION. Let $\mathcal{H}$ be a subset of the space $\mathcal{L}^1(\Omega,\mathcal{F},\mathbb{P})$. $\mathcal{H}$ is called a uniformly integrable set if the integrals 17

(17.1) $$\int_{\{|f| \geq c\}} |f(\omega)|\mathbb{P}(d\omega) \qquad (f \in \mathcal{H})$$

tend uniformly to 0 as the positive number c tends to $+\infty$.

---

(1) For the case of a non-bounded measure, see Dunford-Schwartz [1].

NOTATION. Let f be a random variable. We denote by f^c the function

$$f^c(\omega) = f(\omega) \quad \text{for } |f(\omega)| \leq c$$
$$f^c(\omega) = 0 \quad \text{for } |f(\omega)| > c.$$

We write $f_c = f - f^c$. Definition 17 then takes the following form : $\mathcal{H}$ is uniformly integrable if and only if, for every $\varepsilon > 0$, a number c exists so that $\|f_c\|_1 < \varepsilon$ for every $f \in \mathcal{H}$.

18 REMARKS.

(a) Every family of random variables dominated in absolute value by a fixed integrable function (in particular, every finite subset of $\mathcal{L}^1$) is uniformly integrable.

(b) Definition 17 is obviously compactible with a.s. equality of random variables (1). It only involves the latter through their absolute values ; so we may often restrict ourselves to positive random variables.

19 THEOREM. Let $\mathcal{H}$ be a subset of $\mathcal{L}^1$; for $\mathcal{H}$ to be uniformly integrable, it is necessary and sufficient that the following conditions hold :

(a) the expectations $\mathbb{E}[|f|]$, $f \in \mathcal{H}$, are uniformly bounded (2) ;

(b) for every $\varepsilon > 0$, there exists a number $\delta > 0$ such that the conditions $A \in \mathcal{F}, \mathbb{P}(A) \leq \delta$, imply the inequality

$$\int_A |f(\omega)| \mathbb{P}(d\omega) \leq \varepsilon \quad (f \in \mathcal{H}). \tag{19.1}$$

Proof : To establish the necessity of conditions (a) and (b), we note that, for every integrable function f and every set $A \in \mathcal{F}$,

$$\int_A |f(\omega)| \mathbb{P}(d\omega) \leq c.\mathbb{P}(A) + \mathbb{E}[|f_c|]. \tag{19.2}$$

Suppose that $\mathcal{H}$ is uniformly integrable and choose c so large that

$$\mathbb{E}[|f_c|] < \varepsilon/2 \quad (f \in \mathcal{H}).$$

We first obtain (a) by taking $A = \Omega$, then (b) choosing $\delta = \varepsilon/2c$.

Conversely, suppose that properties (a) and (b) hold, and let $\varepsilon > 0$ be given. Choose some $\delta > 0$ satisfying (b) and let $c = \sup_{f \in \mathcal{H}} \mathbb{E}[|f|]/\delta$, (finite by virtue of (a)). Apply (19.1), taking for A the set $\{|f| \geq c\}$, whose probability is less than δ according to the inequality

$$\mathbb{P}\{|f| \geq c\} \leq \frac{1}{c}\mathbb{E}[|f|] ;$$

we get

$$\int_{\{|f| \geq c\}} |f(\omega)| \mathbb{P}(d\omega) \leq \varepsilon \quad (f \in \mathcal{H})$$

and $\mathcal{H}$ indeed is uniformly integrable.

---

(1) We can thus speak of uniformly integrable subsets of L^1.

(2) It can be proved that (a) is a consequence of (b) if the law $\mathbb{P}$ is diffuse (i.e. has no atomic part).

THEOREM. Let $\mathcal{H}$ be a uniformly integrable set ; the closed convex hull of $\mathcal{H}$ in $\mathcal{L}^1$ is also uniformly integrable. 20

Proof : We begin by noting that the closure of a uniformly integrable set in $\mathcal{L}^1$ is also uniformly integrable : this is an immediate consequence of theorem 19. Hence it suffices to show that the convex hull of $\mathcal{H}$ is uniformly integrable. We check conditions (a) and (b) of 19. The first one is obvious. Let us choose δ such that (19.1) holds for every $f \in \mathcal{H}$; let $f_1, \ldots, f_n$ be elements of $\mathcal{H}$, $t_1, \ldots, t_n$ numbers ≥ 0 such that $t_1 + \ldots + t_n = 1$ and A a measurable set such that $\mathbb{P}(A) \leq \delta$. Then

$$\int_A |t_1 f_1 + \ldots + t_n f_n| \mathbb{P} \leq t_1 \int_A |f_1| \mathbb{P} + \ldots + t_n \int_A |f_n| \mathbb{P} \leq \varepsilon .$$

Hence condition (b) is satisfied.

REMARK. Let H and K be two uniformly integrable subsets of L^1 ; their union $H \cup K$ obviously is uniformly integrable and so is its convex hull; it then follows from the inclusion :

$$\tfrac{1}{2}(H+K) \subset \text{convex hull of } H \cup K.$$

that the sum $H + K$ is uniformly integrable. This result can also be deduced simply from 19.

The following result generalizes the dominated convergence theorem.

THEOREM. Let $(f_n)_{n \in \mathbb{N}}$ be a sequence of integrable random variables which converges almost everywhere [(1)] to a random variable f. Then f is integrable and f_n converges to f in the strong sense in L^1, if and only if the f_n are uniformly integrable. If the random variables f_n are positive, it is also necessary and sufficient that : 21

$$\lim_n \mathbb{E}[f_n] = \mathbb{E}[f] < \infty.$$

Proof : Assume first that the f_n converge to f in L^1 (which supposes the integrability of f) ; we show that conditions (a) and (b) of 19 are satisfied.

We have for $A \in \mathcal{F}$

$$\int_A |f_n(\omega)| \mathbb{P}(d\omega) \leq \int_A |f(\omega)| \mathbb{P}(d\omega) + \|f_n - f\|_1 . \tag{21.1}$$

Condition (a) follows immediately. We choose an integer N such that $\|f_n - f\|_1 \leq \varepsilon/2$ for all $n > N$ and a number δ such that the inequality $\mathbb{P}(A) \leq \delta$ implies $\int_A |g| \mathbb{P} \leq \varepsilon/2$, when g runs through the finite set $\{f_1, f_2, \ldots, f_N, f\}$. The left-hand side of (21.1) is then at most ε for all n provided $\mathbb{P}(A) \leq \delta$, and condition (b) is satisfied.

Conversely, suppose that the functions f_n are uniformly integrable. Then the expectations $\mathbb{E}[|f_n|]$ are uniformly bounded and Fatou's Lemma implies that $\mathbb{E}[|f|] < \infty$. Let us show that f_n converges to f in L^1. We have

$$\mathbb{E}[|f_n - f|] \leq \mathbb{E}[|f_n^c - f^c|] + \mathbb{E}[f_{nc}] + \mathbb{E}[|f_c|] . \tag{21.2}$$

(1) Or only in probability.

Let $\varepsilon > 0$ be given. Choose c so large that the last two expectations are bounded by $\varepsilon/3$ for all n, and such that $\mathbb{P}\{|f| = c\} = 0$ (which is possible, since there are only countably many t such that $\mathbb{P}\{|f| = t\} > 0$. Next we can choose n so large that the first expectation is bounded by $\varepsilon/3$, according to Lebesgue's Theorem, since the functions $|f_n^c - f^c|$ are uniformly bounded and converge almost everywhere to 0. The left-hand side of (21.2) then is at most ε, and convergence in norm is established.

It remains to show that the convergence of $\mathbb{E}[f_n]$ to $\mathbb{E}[f] < \infty$ implies, when the f_n are positive, the convergence of $\mathbb{E}[|f_n - f|]$ to 0 (and consequently the uniform integrability of the f_n). To this end, we write :

$$f + f_n = (f \vee f_n) + (f \wedge f_n).$$

$\mathbb{E}[f \wedge f_n]$ tends to $\mathbb{E}[f]$ by Lebesgue's Theorem. On the other hand, $\mathbb{E}[f + f_n]$ tends to $2\mathbb{E}[f]$ by hypothesis. It follows that $\mathbb{E}[f \vee f_n]$ tends to $\mathbb{E}[f]$. We then deduce from the relation

$$|f - f_n| = f \vee f_n - f \wedge f_n$$

that $\mathbb{E}[|f - f_n|]$ tends to 0.

We give a complete proof of the following theorem (due to la Vallée-Poussin), because it helps to understand the significance of uniform integrability. However, the most useful part of it is the implication (2) $\Rightarrow$ (1), which is also the easier to establish. For example, every bounded subset of L^2 is uniformly integrable (take $G(t) = t^2$).

22 THEOREM. <u>Let $\mathcal{H}$ be a subset of $\mathcal{L}^1$. The following properties are equivalent</u> :

(1) <u>$\mathcal{H}$ is uniformly integrable</u>.

(2) <u>There exists a positive function</u> G(t) <u>defined on</u> $\mathbb{R}_+$ <u>such that</u> $\lim_{t\to\infty} \frac{G(t)}{t} = +\infty$ <u>and</u> (1).

(22.1) $$\sup_{f \in \mathcal{H}} \mathbb{E}[G \circ |f|] < \infty.$$

<u>Proof</u> : The establish that (2) $\Rightarrow$ (1), let $\varepsilon > 0$ be given and let $a = \frac{M}{\varepsilon}$, where M is the value of the left-hand side of (22.1). We choose c so large that $\frac{G(t)}{t} \geq a$ for all $t \geq c$. Then we have $|f| \leq \frac{G \circ |f|}{a}$ on the set $\{|f| \geq c\}$ and consequently

$$\int_{\{|f| \geq c\}} |f|\mathbb{P} \leq \frac{1}{a}\int_{\{|f| \geq c\}} G \circ |f|\mathbb{P} \leq \frac{1}{a} M = \varepsilon$$

for every function $f \in \mathcal{H}$. Definition 17 is therefore satisfied.

We now establish the converse by constructing a function G(t) of the from $\int_0^t g(s)ds$, where g is an increasing function equal to zero at t = 0, which tends to $+\infty$ with t and takes a constant value g_n on each interval $[n, n+1[$ $(n \in \mathbb{N})$. We write, for each function $f \in \mathcal{H}$.

$$a_n(f) = \mathbb{P}\{|f| > n\}.$$

---

(1) The function G which we construct is also convex.

Since $g_0 = 0$, we have
$\mathbb{E}[G \circ |f|] \leq g_1.\mathbb{P}\{1 < |f| \leq 2\} + (g_1+g_2).\mathbb{P}\{2 < |f| \leq 3\} +\ldots = \sum_{n=1}^{\infty} g_n.a_n(f)$.

Hence it remains to show that it is possible to choose coefficients g_n which tend to infinity as n increases, such that the sums $\sum g_n.a_n(f)$ are uniformly bounded. We choose an increasing sequence of integers c_n, which tends to infinity, such that

$$\int_{\{|f| \geq c_n\}} |f|\mathbb{P} \leq 2^{-n} \qquad (f \in \mathcal{H})$$

according to our assumption of uniform integrability. We have :

$$\int_{\{|f| \geq c_n\}} |f|\mathbb{P} \geq \sum_{k=c_n}^{\infty} k\mathbb{P}\{k < |f| \leq k+1\} \geq \sum_{m=c_n}^{\infty} \mathbb{P}\{|f| > m\} = \sum_{m=c_n}^{\infty} a_m(f).$$

It follows that the sum $\sum_n \sum_{c_n}^{\infty} a_m(f)$ is uniformly bounded for $f \in \mathcal{H}$; but this sum is of the form $\sum_m g_m.a_m(f)$, where g_m denotes the number of integers n such that $c_n \leq m$. The theorem is established.

Weak topologies

We now give some results on the weak topology $\sigma(L^1,L^\infty)$ closely related in fact to uniform integrability. We make some use of the conditional expectation operators, which will only be defined later (40), but this involves of course no circularity.

We first recall a well know theorem :

THEOREM (Vitali-Hahn-Saks). Let (μ_n) be a sequence of bounded measures, not necessa- 23
rily positive, on a measurable space $(\Omega,\mathcal{F})$ and let λ be a bounded positive measure such that the μ_n are absolutely continuous with respect to λ. Suppose that for all $A \in \mathcal{F}$ the limit $\mu(A) = \lim_n \mu_n(A)$ exists and is finite. Then

(1) μ is a bounded measure.

(2) For every $\varepsilon > 0$, there exists $\eta > 0$ such that the inequality $\lambda(A) \leq \eta$ implies $\sup_n |\mu_n|(A) \leq \varepsilon$. Further, the masses $\|\mu_n\|$ are uniformly bounded.

Proof : We note first that the existence of λ such that the μ_n are absolutely continuous with respect to λ is not a restriction : it suffices to take $\lambda = \sum|\mu_n|/2^n\|\mu_n\|$. Then comparing (2) and 19, we may state (2) in a different way : the densities μ_n/λ are uniformly integrable with respect to λ .

Let Φ be the subset of $L^1(\lambda)$ consisting of the equivalence classes of indicators of elements of $\mathcal{F}$ (we shall denote these classes by the elements of $\mathcal{F}$ they represent) Φ is closed in L^1, hence Φ is a complete metric space. The functions $A \mapsto \mu_n(A)$ are continuous on Φ and converge pointwise to $A \mapsto \mu(A)$.

Let $\alpha > 0$ and let

$$L_j = \{U \in \Phi : \forall m \geq j, \ \forall n \geq j, \ |\mu_n(U) - \mu_m(U)| \leq \alpha\}.$$

L_j is a closed subset of Φ and the union of the L_j is the whole of Φ. By Baire's Theorem, there exists a j such that L_j has an interior point A. In other words, there exist an integer j and a number $h > 0$ such that the relations

$n \geq j$, $m \geq j$, $\lambda(B \,\Delta\, A) \leq h$ imply $|\mu_n(B) - \mu_n(A)| \leq \alpha$.

Such a j being chosen, let $\eta \in]0,h[$ be such that $\lambda(C) \leq \eta$ implies $|\mu_i(C)| \leq \alpha$ for $i = 0, 1,\ldots, j$ (hence $|\mu_i|(C) \leq 2\alpha$ (1)). For $n \geq j$, we write

$$\begin{aligned}|\mu_n(C)| &\leq |\mu_n(A \cup C) - \mu_n(A)| + |\mu_n(A\backslash C) - \mu_n(A)| \\ &\leq |\mu_n(A \cup C) - \mu_j(A \cup C)| + |\mu_j(A \cup C) - \mu_j(A)| + |\mu_j(A) - \mu_n(A)| \\ &\quad + |\mu_n(A\backslash C) - \mu_j(A\backslash C)| + |\mu_j(A\backslash C) - \mu_j(A)| + |\mu_j(A) - \mu_n(A)| .\end{aligned}$$

Thus $\lambda(C) \leq \eta \Rightarrow \sup_n|\mu_n|(C) \leq 6\alpha$. We deduce the following properties.

(1) Since Ω decomposes into finitely many sets of measure $\leq\eta$ (relative to λ) and finitely many atoms of measure $\geq\eta$, the total masses of the μ_n are bounded. Note that this argument is unnecessary if the μ_n are positive.

(2) Taking $\alpha = \varepsilon/6$, we get the last sentence of the theorem.

(3) The additive set function μ is bounded and property (2) implies that μ is a measure, absolutely continuous with respect to λ. Let indeed (E_k) be a decreasing sequence of elements of $\mathcal{F}$, whose intersection is empty : then $\lambda(E_k) \to 0$ and hence $\mu(E_k) \to 0$; μ is therefore countably additive. We know that μ then is the difference of two positive measures. The theorem is proved.

We now prove a special case, much easier than the general one, of the theorems of Eberlein and Šmulian from the theory of topological linear spaces. As usual we work on a probability space $(\Omega,\mathcal{F},\mathbb{P})$.

24 THEOREM. Let K be a subset of L^1, which is compact under the weak topology $\sigma(L^1,L^\infty)$. If the σ-field $\mathcal{F}$ is separable, K is metrizable. Even if $\mathcal{F}$ is not separable, every sequence of elements of K contains a convergent subsequence.

Proof : Suppose first that $\mathcal{F}$ is separable. Let (H_n) be a sequence of elements of $\mathcal{F}$ which generates it and let $\mathcal{A}$ be the Boolean algebra generated by the H_n ; it is easily verified that $\mathcal{A}$ is countable. On the other hand, if f and g are two elements of L^1, the relation $\int_A f\mathbb{P} = \int_A g\mathbb{P}$ for all $A \in \mathcal{A}$ implies $f = g$ a.s. (cf. I.20). We order the elements of $\mathcal{A}$ into a sequence (A_n) and write, for $f, g \in K$,

$$d(f,g) = \sum a_n^{-1}\left|\int_{A_n} f\mathbb{P} - \int_{A_n} g\mathbb{P}\right| \quad \text{where} \quad a_n = 2^n\left(1+\sup_{h\in K}\left|\int_{A_n} h\mathbb{P}\right|\right).$$

d is a metric on K. The associated topology is Hausdorff and coarser than the (compact) topology of K, and hence is equal to it.

Let (f_n) be a sequence of elements of K and let $\mathcal{F}_0$ be the σ-field generated by the f_n ; $\mathcal{F}_0$ is separable even if $\mathcal{F}$ is not. Let U denote the conditional expectation operator $g \mapsto \mathbb{E}[g|\mathcal{F}_0]$, which maps $L^1(\mathcal{F})$ continuously onto $L^1(\mathcal{F}_0)$; U(K) is a metrizable weakly compact subset of $L^1(\mathcal{F}_0)$ and $U(f_n) = f_n$. Hence we can find a subsequence (f'_n) of the sequence (f_n), which converges to $f \in L^1(\mathcal{F}_0)$ for the topology $\sigma(L^1(\mathcal{F}_0), L^\infty(\mathcal{F}_0))$. Consider now $g \in L^\infty(\mathcal{F})$ and let h be $\mathbb{E}[g|\mathcal{F}_0] \in L^\infty(\mathcal{F}_0)$. We have

$$\int f'_n g\mathbb{P} = \int f'_n h\mathbb{P} \to \int fh\mathbb{P} = \int fg\mathbb{P}.$$

---

(1) Recall that $|\theta|(A) = \sup_{B\subset A}(|\theta(B)| + |\theta(A\backslash B)|)$ for every measure θ.

Thus $f'_n \to f$ relative to the topology $\sigma(L^1(\mathcal{F}), L^\infty(\mathcal{F}))$ and the theorem is established.

REMARK. Theorem 24 extends immediately to all the weak topologies $\sigma(L^p, L^q)$, where q is the conjugate exponent of p and $1 \le p \le \infty$.

The implication (1) $\Rightarrow$ (3) of the next theorem will be a fundamental tool in the following chapters. The other implications will not be used as much, but are still very interesting.

THEOREM (Dunford-Pettis compactness criterion) [1]. Let $\mathcal{H}$ be a subset of the space L^1. The following three properties are equivalent : (25)

(1) $\mathcal{H}$ is uniformly integrable.

(2) $\mathcal{H}$ is relatively compact in L^1 with the weak topology $\sigma(L^1, L^\infty)$.

(3) Every sequence of elements of $\mathcal{H}$ contains a subsequence which converges in the sense of the topology $\sigma(L^1, L^\infty)$.

Proof : We show that (1) $\Rightarrow$ (2). Let $\mathcal{U}$ be an ultrafilter on $\mathcal{H}$; for each function $f \in \mathcal{H}$ and every set $E \in \mathcal{F}$, we define

$$I_f(E) = \int_E f(\omega)\mathbb{P}(d\omega).$$

By the relation $|I_f(E)| \le \mathbb{E}[|f|]$ and condition (a) of 19, the numbers $I_f(E)$ are uniformly bounded. The limit

$$I(E) = \lim_{\mathcal{U}} I_f(E)$$

therefore exists for all $E \in \mathcal{F}$. Clearly the set function $E \mapsto I(E)$ is additive and bounded. By condition (b) of 19 there exists for all $\varepsilon > 0$ a number $\delta > 0$ such that $\mathbb{P}(E) \le \delta$ implies $|I(E)| < \varepsilon$; hence I is a measure which is absolutely continuous with respect to $\mathbb{P}$. There then exists, by the Radon-Nikodym Theorem, a function $\phi \in L^1$ such that for every measurable set E

$$I(E) = \int_E \phi(\omega)\mathbb{P}(d\omega).$$

Assertion (2) will be established if we show that $\mathcal{U}$ converges to ϕ in the weak topology. Obviously

$$\lim_{\mathcal{U}} \mathbb{E}[f.g] = \mathbb{E}[\phi.g]$$

for every function $g \in \mathcal{L}^\infty$ which is a finite linear combination of indicators of sets. Since every function $g \in \mathcal{L}^\infty$ is a uniform limit of such functions, the conclusion follows by uniform convergence.

The assertion (2) $\Rightarrow$ (3) follows from 24. Finally, (3) $\Rightarrow$ (1) : assume indeed that (1) does not hold ; then by (19) $\mathcal{H}$ contains a subsequence (f_n) such that either $\mathbb{E}[|f_n|] \to +\infty$, or there exist a number $\varepsilon > 0$ and n elements A_n of $\mathcal{F}$ such that $\mathbb{P}(A_n) \to 0$ and $\int_A |f_n| \mathbb{P} \ge \varepsilon$. According to 23, (2), this sequence has no weakly convergent subsequence and (3) is false.

The following result illustrates the difference between weak convergence and strong convergence in L^1 : a sequence (f_n) which converges weakly but not strongly

---

(1) For the case of non-bounded measures, see Dunford-Schwartz [1].

oscillates violently around its weak limit.

26 THEOREM. Let (f_n) be a sequence of integrable functions on Ω, which converges to f in the sense of $\sigma(L^1,L^\infty)$. Let $A \in \mathcal{F}$ be such that $f \leq \liminf_n f_n$ a.s. on A ; then $\int_A |f - f_n| \mathbb{P} \to 0$.

Proof : We immediately reduce to the case where $f = 0$ and $A = \Omega$. The functions f_n are uniformly integrable by 25. We choose $\alpha > 0$ such that $\mathbb{P}(U) < \alpha$ implies $\int_U |f_n| \mathbb{P} \leq \varepsilon$ for all n. Then we set, for N an integer,

$$A_n = \{\omega : \inf_{n \geq N} f_n(\omega) \geq -\varepsilon\}$$

and choose N so large that $\mathbb{P}(A_n^c) < \alpha$, according to our hypothesis that $\liminf_n f_n \geq 0$. The sequence (f_n) converges weakly to 0, so we may choose $N' \geq N$ such that $n \geq N'$ implies $|\int_{A_N} f_n \mathbb{P}| \leq \varepsilon$. Then we have, if $n \geq N'$,

$$\int |f_n| \mathbb{P} \leq \int_{A_N} |f_n| \mathbb{P} + \int_{A_N^c} |f_n| \mathbb{P} \leq \int_{A_N} |f_n + \varepsilon| \mathbb{P} + \int_{A_N} \varepsilon \mathbb{P} + \int_{A_N^c} |f_n| \mathbb{P}.$$

The last two integrals are no greater than ε(from the choices of N and α for the second one.) On the other hand, by the definition of A_N, the first integral on the right is equal to $\int_{A_N} (f_n+\varepsilon) \mathbb{P} \leq |\int_{A_N} f_n \mathbb{P}| + \varepsilon \leq 2\varepsilon$ since $n \geq N'$. Hence finally $\int |f_n| \mathbb{P} \leq 4\varepsilon$.

A theorem of Mokobodzki

We have seen earlier (no. 10) that any sequence which converges in probability contains an a.s. convergent subsequence. It sometimes happens (e.g. in the theory of Markov processes) that one is given on some space $(\Omega,\mathcal{F})$ a whole family $(\mathbb{P}_i)$ of probability laws and a sequence (f_n) which converges in probability for each of the $\mathbb{P}_i$. Is it then possible to select a single random variable f such that $f_n \to f$ in probability for each of the P_i ? If we knew how to extract from (f_n) a subsequence (f'_n) converging $\mathbb{P}_i$-a.s. for every i, the function $f = \liminf_n f'_n$ would be the solution. Unfortunately, the procedure in 10 depends on the law $\mathbb{P}_i$.

Mokobodzki has shown that there is a universal extraction procedure (performed by means of a filter, not a subsequence) which yields the existence of f. The proof uses the "continuum hypothesis" or continuum axiom. We shall see later another procedure (1) (also using the continuum hypothesis) which leads to analogous results (cf. Meyer [1]).

27 LEMMA. There exists a filter $\mathcal{U}$ on $\mathbb{N}$ with the following property : for every strictly increasing sequence (s_n) of positive integers, there exists a strictly increasing sequence (t_n) such that

(1) $s_n \leq t_n$ for all sufficiently large n,

(2) for all n, the set $\{t_n, t_{n+1},...\}$ belongs to $\mathcal{U}$.

(1) That of "medial limits", also due to Mokobodzki

Proof : We denote by I the set of all countable ordinals (0.8) and by $\mathcal{S}$ the set of all strictly increasing sequences of positive integers. The continuum axiom affirms the existence of a bijection $i \mapsto s^i$ of I onto $\mathcal{S}$. We shall construct by transfinite induction a mapping $i \mapsto t^i$ of I into $\mathcal{S}$ with the following properties :

(a) $s_n^i \leq t_n^i$ for all sufficiently large n,

(b) if $i < j$, t^j is a subsequence of t^i except for a finite number of terms.

The lemma then follows immediately. For each i let indeed $\mathcal{F}_i$ be the "elementary filter" associated with the sequence t^i, that is, the set of all $A \subset \mathbb{N}$ which contain all but a finite number of the t_n^i ; by property (b), the mapping $i \mapsto \mathcal{F}_i$ is increasing. Hence there exists a filter r containing all the $\mathcal{F}_i$ (even an ultrafilter) and, all the strictly increasing sequences having been enumerated, the filter r satisfies the lemma by virtue of (a).

We pass to the construction. We write $t^0 = s^0$. If t^i is constructed, we take t^{i+1} to be a subsequence of t^i such that $s_k^{i+1} \leq t_k^{i+1}$ for all k (an immediate construction by induction on k). If (a) and (b) hold up to the i-th term, they then hold up to the (i+1)-th term. If i is a limit ordinal and the t_j have been constructed for all $j < i$, we proceed as follows : we choose a strictly increasing sequence of ordinals $j_n < i$ such that $i = \sup_n j_n$. By (b), we can construct sequences u^n by suppressing a finite number of terms at the beginning of t^{j_n} such that u^{n+1} is for all n a subsequence of u^n. We may suppress a few more and assume that $u_0^n \geq s_n^i$ for all n. We then write $t_n^i = u_0^n$; this sequence is a subsequence of each of the sequences t^{j_n}, except for a finite number of terms, and it is by construction "more rapid" than s^i. Hence the induction is possible and the lemma is proved.

We call filters satisfying conditions (1) and (2) of 27 rapid filters.

THEOREM. On a complete probability space $(\Omega, \mathcal{F}, P)$, let (f_n) be a sequence of measurable functions which converges in probability to a function f. Let r be a rapid filter on N? Then for almost all ω 28

$$\lim_r f_n(\omega) = f(\omega).$$

Proof : We reduce it immediately to the case where the f_n and f have values in the interval $[-1,1]$. Then the f_n converge to f in L^1. Let $s = (s_k)$ be a strictly increasing sequence of integers such that $m \geq s_k$ implies $\|f_m - f\| \leq 2^{-k}$, then let $t = (t_k)$ be a sequence such that $s_k \leq t_k$ for all sufficiently large k and r is finer than the elementary filter associated with t (property (2) of 27). Since $\sum \|f_{t_{k+1}} - f_{t_k}\|_1 < \infty$, we a.s. have $\lim f_n(\omega) = f(\omega)$ relative to this elementary filter and a fortiori relative to r, which is finer.

COROLLARY. Let (f_n) be a sequence of $\mathcal{F}$-measurable functions and let $f = \liminf_r f_n$. For every law $\mathbb{P}$ such that the sequence (f_n) converges in $\mathbb{P}$-measure, it can be affirmed that f is equal $\mathbb{P}$-a.s. to an $\mathcal{F}$-measurable function and that $f_n \to f$ in $\mathbb{P}$-measure. 29

Note however that the limit f is not universally measurable in general. The similar procedure, of "medial limits" , always leads to universally measurable functions.

3. COMPLETION. INDEPENDENCE. CONDITIONING

We now come back to elementary results of a probabilistic nature.

Internally negligible sets

30 DEFINITION. Let $(\Omega,\mathcal{F},\mathbb{P})$ be a probability space. A set $A \subset \Omega$ is called internally $\mathbb{P}$-negligible if $\mathbb{P}(B) = 0$ for every $B \in \mathcal{F}$ contained in A.

31 THEOREM. Let $\mathcal{N}$ be a family of subsets of Ω which satisfies the following conditions:

(1) $\mathcal{N}$ is closed under $(\cup c)$.

(2) Every element of $\mathcal{N}$ is internally $\mathbb{P}$-negligible.

Let $\mathcal{F}'$ be the σ-field generated by $\mathcal{F}$ and $\mathcal{N}$. The law $\mathbb{P}$ then can be extended uniquely to a law $\mathbb{P}'$ on $\mathcal{F}'$ such that every element of $\mathcal{N}$ is $\mathbb{P}'$-negligible.

Proof : We merely indicate the main steps, leaving details to the reader.

Let $\mathcal{M}$ be the family of subsets of Ω which are contained in some element of $\mathcal{N}$ and $\mathcal{G}$ be the family of subsets of the form $F \Delta M$ ($F \in \mathcal{F}$, $M \in \mathcal{M}$). One checks easily that $\mathcal{G}$ is a σ-field ; since $\mathcal{M}$ contains the empty set, $\mathcal{F} \subset \mathcal{G}$ and similarily $\mathcal{M} \subset \mathcal{G}$.

Let $A = F \Delta M$ be an element of $\mathcal{G}$; we set $\mathbb{Q}(A) = \mathbb{P}(F)$. It can be verified that $\mathbb{Q}(A)$ depends only on A, not on the representation $F \Delta M$ of A.

To show that $\mathbb{Q}$ is a probability law on $\mathcal{G}$, we consider a sequence (A_n).of disjoint elements of $\mathcal{G}$ and their union A. Each A_n is of the form $F_n \Delta M_n$ ($F_n \in \mathcal{F}$, $M_n \in \mathcal{M}$). Let F be the union of the F_n. Since the F_n are disjoint up to a negligible set, $\mathbb{P}(F) = \sum_n \mathbb{P}(E_n)$; on the other hand, A and F differ only by an element of $\mathcal{M}$. Hence $\mathbb{Q}(A) = \sum_n \mathbb{Q}(A_n)$. The required law $\mathbb{P}'$ is then the restriction of $\mathbb{Q}$ to $\mathcal{F}'$.

To establish uniqueness, we consider another law $\mathbb{P}''$ on $\mathcal{F}'$ extending $\mathbb{P}$, such that every element of $\mathcal{N}$ is $\mathbb{P}''$-negligible. Every element of $\mathcal{M}$ then is internally $\mathbb{P}''$-negligible, so that P" extends to a law on $\mathcal{G}$ such that every element of $\mathcal{M}$ is negligible. This law is then identical to $\mathbb{Q}$ and hence $\mathbb{P}' = \mathbb{P}''$.

32 REMARKS.

(a) Theorem 31 is often applied to a family $\mathcal{N}$ consisting of a single internally negligible set.

(b) The theorem implies the possibility of completing (cf. 3) a probability space $(\Omega,\mathcal{F},\mathbb{P})$: one takes for $\mathcal{N}$ the class of all subsets of $\mathbb{P}$-negligible sets.

Let then $\mathcal{F}^{\mathbb{P}}$ be the completed σ-field ; every element of $\mathcal{F}^{\mathbb{P}}$ can be expressed as $F \Delta M$, where F belongs to $\mathcal{F}$ and M is contained in some $\mathbb{P}$-negligible set $N \in \mathcal{F}$. Then $F \Delta M$ lies between the two sets $F\setminus N$ and $F \cup N$, which belong to $\mathcal{F}$ and differ only by a negligible set, and this property obviously characterizes the elements of

$\mathcal{F}^{\mathbb{P}}$. The usual approxiamation of measurable functions by step functions now gives the following result :

A real-valued function f is measurable relative to the completed σ-field $\mathcal{F}^{\mathbb{P}}$, if and only if there exist two $\mathcal{F}$-measurable real-valued functions g and h such that

$$g \le f \le h, \ \mathbb{P}\{g \neq h\} = 0.$$

(c) Let $(\Omega,\mathcal{F})$ be a measurable space ; for each law $\mathbb{P}$ on $(\Omega,\mathcal{F})$ consider the completed σ-field $\mathcal{F}^{\mathbb{P}}$ and denote by $\hat{\mathcal{F}}$ the intersection of all the σ-fields $\mathcal{F}^{\mathbb{P}}$: the measurable space $(\Omega,\hat{\mathcal{F}})$ is called the universal completion of $(\Omega,\mathcal{F})$. The reader can verify the following properties :

(1) Every law $\mathbb{P}$ on $\mathcal{F}$ can be extended uniquely to a law $\hat{\mathbb{P}}$ on $\hat{\mathcal{F}}$ and the mapping $\mathbb{P} \mapsto \hat{\mathbb{P}}$ is a bijection of the set of laws on $\mathcal{F}$ onto the set of laws on $\hat{\mathcal{F}}$.

(2) Let $(E,\mathcal{E})$ be a measurable space and f a measurable mapping of $(\Omega,\mathcal{F})$ into $(E,\mathcal{E})$; then f is a measurable mapping of $(\Omega,\hat{\mathcal{F}})$ into $(E,\hat{\mathcal{E}})$.

(d) The universal completion of a Borel σ-field $\mathcal{B}(E)$ is denoted by $\mathcal{B}_u(E)$ and is also called the σ-field of universally measurable sets of E. If E and F are two topological spaces and f is a mapping from E to F, f is called universally measurable if it is measurable from $\mathcal{B}_u(E)$ to $\mathcal{B}_u(F)$. By (c) it suffices that it be measurable from $\mathcal{B}_u(E)$ to $\mathcal{B}(F)$.

Independence

Text books on elementary probability theory give an important place to independence. In this book, we shall not need it much and we refer the reader to Chung [1] for a more detailed study.

DEFINITION. Let $(X_i)_{i \in I}$ be a finite family of random variables from a probability space $(\Omega,\mathcal{F},\mathbb{P})$ to measurable spaces $(E_i,\mathcal{E}_i)_{i \in I}$. Let X be the random variable $(X_i)_{i \in I}$ with values in the space $(\prod_{i\in I} E_i, \prod_{i\in I} \mathcal{E}_i)$. The random variables X_i (or the family (X_i)) are said to be independent if the law of X is the product of the laws of the X_i. 33

Let $(X_i)_{i \in I}$ be an arbitrary family of random variables. The family (X_i) is said to be independent if every finite subfamily is independent.

More concretely (cf. 14) : the random variables $(X_i)_{i \in I}$ are independent if and only if

$$\mathbb{P}\{\forall i \in J, X_i \in A_i\} = \prod_{i\in J} \mathbb{P}\{X_i \in A_i\}$$

for every finite subset $J \subset I$ and every family $(A_i)_{i \in J}$ such that $A_i \in \mathcal{E}_i$ for all $i \in J$.

The definition of independence can be given another form.

DEFINITION. Let $(\Omega,\mathcal{F},\mathbb{P})$ be a probability space and let $(\mathcal{F}_i)_{i \in I}$ be a family of sub-σ-fields of $\mathcal{F}$. The σ-fields $(\mathcal{F}_i)_{i \in I}$ are called independent if 34

$$\mathbb{P}(\bigcap_{i\in J} A_i) = \prod_{i\in J} \mathbb{P}(A_i)$$

<u>for every finite subset</u> $J \subset I$ <u>and every family of sets</u> $(A_i)_{i \in J}$ <u>such that</u> $A_i \in \mathcal{F}_i$ <u>for all</u> $i \in J$.

The definitions 33 and 34 can easily be reduced to each other. The random variables $(X_i)_{i \in J}$ are indeed independent (in the sense of 33) if and only if the σ-fields $\mathcal{F}(X_i)$ are independent (in the sense of 34). Similarly, the σ-fields $(\mathcal{F}_i)_{i \in J}$ are independent if and only if the random variables X_i are independent, X_i denoting the identity mapping from $(\Omega,\mathcal{F})$ to $(\Omega,\mathcal{F}_i)$.

35 THEOREM. <u>Let</u> $\mathcal{F}_1, \mathcal{F}_2, \ldots, \mathcal{F}_n$ <u>be independent</u> σ-<u>fields and let</u> $f_1, f_2, \ldots, f_n$ <u>be integrable real-valued random variables, measurable relative to the corresponding</u> σ-<u>fields</u> $\mathcal{F}_1, \mathcal{F}_2, \ldots, \mathcal{F}_n$. <u>Then the product</u> $f_1 f_2 \ldots f_n$ <u>is integrable and</u>

$$\mathbb{E}[f_1.f_2\ldots f_n] = \mathbb{E}[f_1].\mathbb{E}[f_2] \ldots \mathbb{E}[f_n].$$

<u>Conditioning</u>

The notion of conditional expectation is essential to probability theory. We give the different forms of the definitions in nos. 36-39 and then, in no. 40, we list all properties that must be kept in mind.

36 THEOREM. <u>Let</u> $(\Omega,\mathcal{F},\mathbb{P})$ <u>be a probability space and</u> f <u>be a random variable from</u> $(\Omega,\mathcal{F})$ <u>to some measurable space</u> $(E,\mathcal{E})$. <u>Let</u> $\mathbb{Q}$ <u>be the image law of</u> $\mathbb{P}$ <u>under</u> f.

<u>Let</u> X <u>be a</u> $\mathbb{P}$-<u>integrable random variable on</u> $(\Omega,\mathcal{F})$. <u>There exists a</u> $\mathbb{Q}$-<u>integrable random variable</u> Y <u>on</u> $(E,\mathcal{E})$ <u>such that, for every set</u> $A \in \mathcal{E}$:

(36.1) $$\int_A Y(x)\mathbb{Q}(dx) = \int_{f^{-1}(A)} X(\omega)\mathbb{P}(d\omega).$$

<u>If</u> Y' <u>is any random variable satisfying</u> (36.1), <u>then</u> $Y' = Y$ <u>a.s.</u>

<u>Proof</u> : The assertion concerning uniqueness of Y is an immediate consequence of remark 9, (a).

To establish the existence of Y, we begin by assuming that X belongs to $\mathcal{L}^2(P)$. We associate to every $Z \in \mathcal{L}^2(\mathbb{Q})$ the number $\int_\Omega (Z \circ f)X.\mathbb{P}$, which depends only on the equivalence class of Z. We thus get a linear functional on $\mathcal{L}^2(\mathbb{Q})$, whose norm is at most $\|X\|_2$. Hence there exists a function $Y \in \mathcal{L}^2(\mathbb{Q})$ such that

$$\int_\Omega (Z \circ f)X.\mathbb{P} = \int_E ZY.Q \qquad (Z \in \mathcal{L}^2(\mathbb{Q})).$$

The function Y solves the problem.

If further X is positive, Y has a positive integral on every set $A \in \mathcal{E}$; hence it is a.s. positive by 9, (a).

We now pass to the case where X is only integrable. The same is true of its positive part X^+ and its negative part X^-. The random variables $X_n^+ = X^+ \wedge n$ $(n \in \mathbb{N})$ belong to $\mathcal{L}^2(\mathbb{P})$. Hence we can associate to them random variables Y_n^+ as above. By the preceding remark, these random variables are a.s. positive and increase with n a.s. and their integrals are bounded by $\mathbb{E}[X^+]$. Hence we can choose an integrable random variable $\overset{+}{Y}$, a.s. equal to the limit of the $\overset{+}{Y}_n$. Similarly we construct form X^- a random variable $\bar{Y}$; the integrable random variable $Y = \overset{+}{Y} - \bar{Y}$ satisfies (36.1) and the theorem is established.

DEFINITION. Let Y be an $\mathcal{E}$-measurable and $\mathbb{Q}$-integrable random variable satisfying relation (36.1). We call Y (a version of) the conditional expectation of X given f. 37

This will be denoted provisionally by $\mathbb{E}[X/f]$; this notation will not be used after 39.

REMARKS. 38

(a) If X is the indicator of an event B, $\mathbb{E}[X/f]$ is called the conditional probability of B, given f. It is important to keep in mind that such a "probability" is not a number, but a random variable defined up to equivalence.

(b) Consider a partition of the set Ω into a sequence of measurable sets A_n and denote by f the mapping of Ω into $\mathbb{N}$ equal to n on A_n. A measure $\mathbb{Q}$ on $\mathbb{N}$ then is defined by :

$$\mathbb{Q}(\{n\}) = \mathbb{P}(A_n).$$

Let X be an integrable random variable on Ω; it is easy to compute $Y = \mathbb{E}[X/f]$

$$Y(n) = \frac{\int_{A_n} X.\mathbb{P}}{\mathbb{P}(A_n)} \quad \text{for all } n \text{ such that } \mathbb{P}(A_n) \neq 0.$$

If $\mathbb{P}(A_n)$ is zero, Y(n) can be chosen arbitrarily (1). Suppose in particular that X is the indicator of an event B ; then $Y(n) = \frac{\mathbb{P}(B \cap A_n)}{\mathbb{P}(A_n)}$ if $\mathbb{P}(A_n)$ is non-zero. We recognize here the number which is called, in elementary probability theory, the conditional probability of B given that A_n occur. It would be tempting to use the same terminology in the general case and to call the value Y(x) ($x \in E$) "the conditional expectation of X given that $f(\omega) = x$", but this would be improper, since the random variable Y is only defined up to $\mathbb{Q}$-equivalence and one may not talk about its value at a point x unless $\mathbb{Q}(\{x\}) \neq 0$.

(a) Let X be a non-integrable positive random variable. Passing to the monotone limit as in the proof of 36 gives a positive random variable Y, finite or not, defined up to a.s. equality, which satisfies formula (36.1). We still denote it by $\mathbb{E}[X/f]$ and speak in this case of a generalized conditional expectation. Then $\mathbb{E}[X/f]$ is finite a.s. if and only if there exists an increasing sequence (A_n) of elements of $\mathcal{E}$, such that 39

$$\bigcup_n A_n = E, \quad \int_{f^{-1}(A_n)} X\mathbb{P} < +\infty \quad \text{for all } n.$$

(b) Given an arbitrary random variable X, we now say that X has a generalized conditional expectation if $\mathbb{E}[X^+/f]$ and $\mathbb{E}[X^-/f]$ are finite a.s. and we then set $\mathbb{E}[X/f] = \mathbb{E}[X^+/f] - \mathbb{E}[X^-/f]$.

We started with definition 37 of conditional expectations, because it may be the most intuitive one. But it has a for more important variant, in fact, the only

(1) Usually we take Y(n) = 0, in conformity with the convention $\frac{0}{0} = 0$.

form that we shall use hence forth. One gets it by taking, in statements 37-39, E to be Ω, $\mathcal{E}$ to be a sub-σ-field of $\mathcal{F}$ and f to be the identity mapping. The image measure $\mathbb{Q}$ then is the restriction of $\mathbb{P}$ to $\mathcal{E}$ and we have the following definition :

40 DEFINITION. Let $(\Omega,\mathcal{F},\mathbb{P})$ be a probability space, $\mathcal{E}$ be a sub-σ-field of $\mathcal{F}$ and X be an integrable random variable. A (version of the) conditional expectation of X given $\mathcal{E}$ is any $\mathcal{E}$-measurable integrable random variable Y such that

(40.1) $$\int_A X(\omega)\mathbb{P}(d\omega) = \int_A Y(\omega)\mathbb{P}(d\omega) \quad \text{for all } A \in \mathcal{E}.$$

In general we omit the word "version". We denote Y be the notation $\mathbb{E}[X|\mathcal{E}]$ (1). If $\mathcal{E}$ is the σ-field $\sigma(f_i, i \in I)$ generated by a family of random variables, we speak of the conditional expectation of X given the f_i and write $\mathbb{E}[X|f_i, i \in I]$. If X is the indicator of an event A, we speak of the conditional probability of A given $\mathcal{E}$ (or the f_i) and write $\mathbb{P}(A|\mathcal{E})$, $\mathbb{P}(A|f_i, i \in I)$. It often happens that conditional expectations are iterated as in

$$\mathbb{E}[\mathbb{E}[X|\mathcal{F}_1]|\mathcal{F}_2] ,$$

$\mathcal{F}_1$ and $\mathcal{F}_2$ being two sub-σ-fields of $\mathcal{F}$. We then use the simpler notation $\mathbb{E}[X|\mathcal{F}_1|\mathcal{F}_2]$, which is entirely unambiguous.

REMARKS.

(a) Coming back to the notation of 36-37, denote by $\mathcal{S}$ the σ-field $\sigma(f)$; we have the a.s. equality $\mathbb{E}[X|\mathcal{S}] = Y \circ f$. Theorem I.18 then reduces Definition 37 to Definition 40.

(b) A random variable X (not assumed to be positive or integrable) has a generalized conditional expectation given $\mathcal{E}$ if and only if the measure $|X|.\mathbb{P}$ is σ-finite on $\mathcal{E}$.

Fundamental properties of conditional expectations

(41) We group under this heading all the properties of conditional expectations which we use later on. In particular, we state again Definition 40 in another way.

All random variables concerned are defined on $(\Omega,\mathcal{F},\mathbb{P})$.

PROPERTY 1. Let X and Y be integrable random variables and a, b, c be constants. Then, for every σ-field $\mathcal{E}\subset\mathcal{F}$,

(41.1) $$\mathbb{E}[aX + bY + c|\mathcal{E}] = a.\mathbb{E}[X|\mathcal{E}] + b.\mathbb{E}[Y|\mathcal{E}] + c \quad \text{a.s.}$$

PROPERTY 2. Let X and Y be integrable random variables such that $X \leq Y$ a.s. Then $\mathbb{E}[X|\mathcal{E}] \leq \mathbb{E}[Y|\mathcal{E}]$ a.s.

PROPERTY 3. Let X_n ($n \in \mathbb{N}$ be integrable random variables which increase to an integrable random variable X. Then

(41.2) $$\mathbb{E}[X|\mathcal{E}] = \lim_n \mathbb{E}[X_n|\mathcal{E}] \quad \text{a.s.}$$

---

(1) Hunt simply writes $\mathcal{E}X$. This is an excellent notation !

PROPERTY 4. (Jensen's inequality). *Let* c *be a convex mapping of* $\mathbb{R}$ *into* $\mathbb{R}$ *and let* X *be an integrable random variable such that* $c \circ X$ *is integrable. We then have*

$$c \circ \mathbb{E}[X|\mathcal{E}] \leq \mathbb{E}[c \circ X|\mathcal{E}] \text{ a.s.} \tag{41.3}$$

Proof : The function c is the upper envelope of a countable family of affine functions $L_n(x) = a_n x + b_n$. The random variables $L_n \circ X$ are integrable and

$$L_n \circ \mathbb{E}[X|\mathcal{E}] = \mathbb{E}[L_n \circ X|\mathcal{E}] \leq \mathbb{E}[c \circ X|\mathcal{E}].$$

Then we take the upper envelope on the right-hand side. If X takes its values in some interval I of $\mathbb{R}$, it obviously suffices that c be convex on I.

PROPERTY 5. *Let* X *be an integrable random variable* ; *then* $\mathbb{E}[X|\mathcal{E}]$ *is* $\mathcal{E}$-*measurable* ; *if* X *is* $\mathcal{E}$-*measurable, then* $X = \mathbb{E}[X|\mathcal{E}]$ *a.s.*
(This is a partial restatement of the definition of conditional expectations; with an obvious consequence of their uniqueness).

PROPERTY 6. *Let* $\mathcal{D}$, $\mathcal{E}$, *be two sub*-σ-*fields of* $\mathcal{F}$ *such that* $\mathcal{D} \subset \mathcal{E}$. *Then for every integrable random variable* X

$$\mathbb{E}[X|\mathcal{E}|\mathcal{D}] = \mathbb{E}[X|\mathcal{D}] \text{ a.s.} \tag{41.4}$$

And in particular

$$\mathbb{E}[\mathbb{E}[X|\mathcal{E}]] = \mathbb{E}[X]. \tag{41.5}$$

(The first formula is an immediate consequence of uniqueness. The second follows by taking $\mathcal{D} = \{\emptyset, \Omega\}$.)

PROPERTY 7. *Let* X *be an integrable random variable and* Y *be an* $\mathcal{E}$-*measurable random variable such that* XY *is integrable. Then*

$$\mathbb{E}[XY|\mathcal{E}] = Y.\mathbb{E}[X|\mathcal{E}] \text{ a.s.} \tag{41.6}$$

Proof : When Y assumes only finitely many values, (41.6) is an immediate consequence of the definition of conditional expectations. The general case follows by monotone convergence.

The extension of these properties to generalized conditional expectations is sometimes useful. We leave it to the reader.

CONTINUITY PROPERTIES 42

We apply Jensen's inequality taking c(x) to be the function $|x|^p$ $(1 \leq p \leq \infty)$. We get

$$\| \mathbb{E}[X|\mathcal{E}] \|_p \leq \| X \|_p . \tag{42.1}$$

The same inequality is obvious for $p = \infty$. The mapping $X \mapsto \mathbb{E}[X|\mathcal{E}]$ therefore is an operator of norm ≤ 1 on L^p $(1 \leq p \leq \infty)$. Now it is well known that a continuous linear operator on a Banach space B still is continuous when B is given its weak topology $\sigma(B, B^*)$ see for example Bourbaki [1] (1), Dunford-Schwartz [1], p. 422).

(1) E.V.T. IV, 2nd edition, §4, no. 2, Proposition 6 (page 103).

Hence the conditional expectation operators are continuous for the weak topologies $\sigma(L^1,L^\infty)$ and $\sigma(L^2,L^2)$, for example.

> Let $(X_n)_{n\in\mathbb{N}}$ be a sequence of integrable random variables which converges a.s. to an integrable random variable X. It may be asked whether the conditional expectations $\mathbb{E}[X_n|\mathcal{E}]$ converge a.s. to $\mathbb{E}[X|\mathcal{E}]$, for any σ-field $\mathcal{E}$. Doob has shown that the answer is yes if the X_n are dominated by a fixed integrable function, and Blackwell and Dubins have shown in [1] that this condition cannot be improved.

Conditional independence

The proof of Theorem 45 may be a good exercise on Properties 1-7 above.

43 DEFINITION. *Let* $(\Omega,\mathcal{F},\mathbb{P})$ *be a probability space and* $\mathcal{F}_1,\mathcal{F}_2,\mathcal{F}_3$ *be three sub-*σ*-fields of* $\mathcal{F}$. $\mathcal{F}_1$ *and* $\mathcal{F}_3$ *are called* conditionally independent given $\mathcal{F}_2$ *if*

(43.1) $$\mathbb{E}[Y_1Y_3|\mathcal{F}_2] = \mathbb{E}[Y_1|\mathcal{F}_2].\mathbb{E}[Y_3|\mathcal{F}_2] \text{ a.s.}$$

where Y_1, Y_3 *denote positive random variables measurable with respect to the corresponding* σ*-field* $\mathcal{F}_1$, $\mathcal{F}_3$.

44 REMARKS. (a) Taking $\mathcal{F}_2$ to be the σ-field $\{\emptyset,\Omega\}$, we recover the definition of independence (33,34). We could similarily define conditional independence of several σ-fields relative to a given σ-field.

(b) It can easily be shown, through the usual monotone limit procedure, that it suffices to assume (43.1) when Y_1 and Y_3 are indicators of sets.

45 THEOREM. *Let* $\mathcal{F}_{12}$ *be the* σ*-field generated by* $\mathcal{F}_1$ *and* $\mathcal{F}_2$. *Then* $\mathcal{F}_1$ *and* $\mathcal{F}_3$ *are conditionally independent given* $\mathcal{F}_2$, *if and only if*

(45.1) $$\mathbb{E}[Y_3|\mathcal{F}_{12}] = \mathbb{E}[Y_3|\mathcal{F}_2] \text{ a.s.}$$

for every $\mathcal{F}_3$ *measurable and integrable random variable* Y_3.

Proof : (a) (43.1) $\Rightarrow$ (45.1). We wish to check that both sides of (45.1) have the same integral on every element of $\mathcal{F}_{12}$. Now the set of elements of $\mathcal{F}_{12}$ for which this property holds is closed under ($\cup$mc,$\cap$mc). On the other hand, the family $\mathcal{C}$ of finite unions of disjoint sets of the form $A_1 \cap A_2$ ($A_1 \in \mathcal{F}_1$, $A_2 \in \mathcal{F}_2$) generates $\mathcal{F}_{12}$. Hence it suffices, by I.19, to verify that

$$\mathbb{E}[a_1a_2.\mathbb{E}[Y_2|\mathcal{F}_{12}]] = \mathbb{E}[a_1a_2.\mathbb{E}[Y_3|\mathcal{F}_2]]$$

where a_1 and a_2 denote respectively the indicators of A_1 and A_2. Now we have (the numbers indicating the properties used) :

$$\mathbb{E}[a_1a_2.\mathbb{E}[Y_3|\mathcal{F}_{12}]] = \mathbb{E}[\mathbb{E}[a_1a_2Y_3|\mathcal{F}_{12}]] \qquad (7)$$

$$= \mathbb{E}[a_1a_2Y_3] \qquad (5)$$

$$= \mathbb{E}[\mathbb{E}[a_1a_2Y_3|\mathcal{F}_2]] \qquad (5)$$

$$= \mathbb{E}[a_2.\mathbb{E}[a_1Y_3|\mathcal{F}_2]] \qquad (7)$$

$$= \mathbb{E}[a_2.\mathbb{E}[a_1|\mathcal{F}_2].\mathbb{E}[Y_3|\mathcal{F}_2]] \qquad (43.1)$$

$$= \mathbb{E}[a_2 . \mathbb{E}[(a_1 . \mathbb{E}[Y_3|\mathcal{F}_2])|\mathcal{F}_2]] \qquad (7)$$

$$= \mathbb{E}[\mathbb{E}[(a_2 a_1 \mathbb{E}[Y_3|\mathcal{F}_2])|\mathcal{F}_2]] \qquad (7)$$

$$= \mathbb{E}[a_2 a_1 \mathbb{E}[Y_3|\mathcal{F}_2]] \qquad (5)$$

(b) (45.1) $\Rightarrow$ (43.1). We have :

$$\mathbb{E}[Y_1 Y_2|\mathcal{F}_2] = \mathbb{E}[Y_1 Y_3|\mathcal{F}_{12}|\mathcal{F}_2] \qquad (6)$$

$$= \mathbb{E}[(Y_1 . \mathbb{E}[Y_3|\mathcal{F}_{12}])|\mathcal{F}_2] \qquad (7)$$

$$= \mathbb{E}[(Y_1 . \mathbb{E}[Y_3|\mathcal{F}_2])|\mathcal{F}_2] \qquad (45.1)$$

$$= \mathbb{E}[Y_1|\mathcal{F}_2] . \mathbb{E}[Y_3|\mathcal{F}_2] \qquad (7)$$

CHAPTER III

Complements to measure theory

Thanks to Hunt [1], Choquet's theorem on capacitability has become one of the fundamental tools of probability theory. This theorem is proved in paragraph 2 and constitutes the core of the chapter. Paragraph 1 contains the elements of analytic set theory necessary to prove Choquet's theorem and other results useful to probabilists (Blackwell's theorem for instance). Paragraph 3 is devoted to bounded Radon measures.

We have tried to restrict ourselves to really useful results, either for probability theory or for potential theory except in (the appendix, which contains some luxury theorems). But this does not mean they are all equally important. The reader that looks for essentials may limit himself to nos 1-13, 27-32 and 44.

1. ANALYTIC SETS

Let E be a set. A <u>paving</u> on E is any family of subsets of E which contains the empty 1
set ; the pair $(E,\mathcal{E})$ consisting of a set E and a paving $\mathcal{E}$ on E is called a <u>paved set</u>. This terminology is used only in this chapter and the applications which depend on it.

Let $(E_i,\mathcal{E}_i)_{i \in I}$ be a family of paved sets. The <u>product paving</u> of the $\mathcal{E}_i$ (resp. the <u>sum paving</u> (1) of the $\mathcal{E}_i$) is the paving on the set $\prod_{i\in I} E_i$ (resp. $\sum_{i\in I} E_i$) consisting of the subsets of the form $\prod_{i\in I} A_i$ (resp. $\sum_{i\in I} A_i$) where $A_i \subset E_i$ belongs to $\mathcal{E}_i$ for all i (and, in the case of the sum, differs from $\emptyset$ only for finitely many indices).

> The first edition of this book gave a different definition of the product paving, analogous to that of the sum paving, insisting that $A_i = E_i$ except for a finite number of indices. It then follows, when this number is equal to 0, that the whole space belongs to every product paving, which causes some inconveniences. The present definition is better, given that we only consider <u>countable</u> products (or sums)

---

(1) Recall that the sum of the E_i (denoted by $\sum_{i\in I} E_i$ or $\coprod_{i\in I} E_i$) is the union of the sets $E_i \times \{i\}$.

It should be noted that, when the $\mathcal{E}_i$ are σ-fields, the product paving of the $\mathcal{E}_i$ is not the same as the product σ-field of the $\mathcal{E}_i$ (the latter being generated by the product paving when I is countable). Hence there is some ambiguity in using notations such as $\prod_{i\in I} \mathcal{E}_i$ or $\mathcal{E} \times \mathcal{F}$ to denote a product paving. We shall nevertheless use them, in this chapter only (1), being explicit when necessary.

Compact and semi-compact pavings

2 Let $(E,\mathcal{E})$ be a paved set and $(K_i)_{i \in I}$ be a family of elements of $\mathcal{E}$. We say that this family has the finite intersection property if $\bigcap_{i\in I_0} K_i \neq \emptyset$ for every finite subset $I_0 \subset I$. This amounts to saying that the sets K_i belong to a filter or also, by the ultrafilter theorem (2), that there exists an ultrafilter $\mathcal{U}$ such that $K_i \in \mathcal{U}$ for all $i \in I$.

3 DEFINITION. Let $(E,\mathcal{E})$ be a paved set. The paving $\mathcal{E}$ is said to be compact (resp. semi-compact) if every family (resp. every countable family) of elements of $\mathcal{E}$, which has the finite intersection property, has a non-empty intersection (3).

For instance, if E is a Hausdorff topological space, the paving consisting of the compact subsets of E (henceforth denoted by $\mathcal{K}(E)$) is compact. Abstract compact pavings are seldom found : an interesting example is that of the "islets" of $\mathbb{N}^{\mathbb{N}}$ (no 77 in the appendix).

Let $\mathcal{E}$ be a compact (resp. semi-compact) paving on E ; then the paving $\mathcal{E} \cup \{E\}$ is compact (resp. semi-compact).

> The definition of analytic sets given in this edition no longer uses semi-compact pavings. Hence the reader can omit every reference to it. The reasons for retaining it are of a purely aesthetic nature.

4 THEOREM. Let E be a set with a compact (resp. semi-compact) paving $\mathcal{E}$ and let $\mathcal{E}'$ be the closure of $\mathcal{E}$ under $(\cup f,\cap a)$ (resp. $(\cup f,\cap c)$). Then the paving $\mathcal{E}'$ is compact (resp. semi-compact).

Proof : Let $\mathcal{F}$ be the closure of $\mathcal{E}$ under $(\cup f)$. Then $\mathcal{E}'$ is the closure of $\mathcal{F}$ under $(\cap a)$ (resp. $(\cap c)$). The latter closure obviously preserves compactness and hence it will suffice to show that $\mathcal{F}$ is compact (resp. semi-compact). So let us consider a family (resp. a countable family) $(K_i)_{i \in I}$ of elements of $\mathcal{F}$, which has the finite intersection property ; let $\mathcal{U}$ be an ultrafilter such that $K_i \in \mathcal{U}$ for all i. Each set K_i is a union $\bigcup_{j\in J_i} K_{ij}$ of elements of $\mathcal{E}$, where J_i is a finite set. Hence there exists

(1) The best solution consists in using the symbol ⊗ for product σ-fields, as does Neveu [1].

(2) Bourbaki [2] (3rd edition), §6 no 4, Theorem 1.

(3) Here is a simple example of a non-compact semi-compact paving : on a non-countable set, the paving consisting of all finite subsets and all subsets whith a countable complement.

an index $j_i \in J_i$ such that $K_{ij_i} \in \mathcal{U}$ (1). The family $(K_{ij_i})_{i \in I}$ therefore has the finite intersection property, hence its intersection is non-empty and so a fortiori is the intersection of the family $(K_i)_{i \in I}$.

THEOREM. Let $(E_i, \mathcal{E}_i)_{i \in I}$ be a family of paved sets. If each of the pavings $\mathcal{E}_i$ is compact (resp. semi-compact) so are the product paving $\prod_{i \in I} \mathcal{E}_i$ and the sum paving $\sum_{i \in I} \mathcal{E}_i$. 5

Proof : The proof is immediate as far as the product paving is concerned. Let $\mathcal{H}$ be the paving on the sum set $\sum_{i \in I} E_i$ consisting of all subsets of the form $\sum_{i \in I} A_i$ such that $A_i = \emptyset$ for all indices except at most one i, for which A_i belongs to $\mathcal{E}_i$. This paving is obviously compact (semi-compact). It then suffices to note that the sum paving is the closure of $\mathcal{H}$ under $(\cup f)$.

There is no need to attach any importance to the "semi"-compact nature of the paving in the following statement : the gain in generality is illusory.

THEOREM. Let $(E, \mathcal{E})$ be a paved set and let f be a mapping of E into a set F. Suppose that, for all $x \in F$, the paving consisting of the sets $f^{-1}(\{x\}) \cap A$, $A \in \mathcal{E}$, is semi-compact. Then, for every decreasing sequence $(A_n)_{n \in \mathbb{N}}$ of elements of $\mathcal{E}$. 6

(6.1) $$f\left(\bigcap_{n \in \mathbb{N}} A_n\right) = \bigcap_{n \in \mathbb{N}} f(A_n).$$

Proof : It suffices to show that we can associate to every $x \in \bigcap_n f(A_n)$ an element $y \in \bigcap_n A_n$ such that $f(y) = x$. Now the family of sets of the form $f^{-1}(\{x\}) \cap A_n$ has the finite intersection property, hence it has a non-empty intersection and we just choose y in this intersection.

$\mathcal{F}$-analytic sets

DEFINITION. Let $(F, \mathcal{F})$ be a paved set. A subset A of F is called $\mathcal{F}$-analytic if there exist an auxiliary compact metrizable space E and a subset $B \subset E \times F$ belonging to $(\mathcal{K}(E) \times \mathcal{F})_{\sigma\delta}$, such that A is the projection of B onto F. The paving on F consisting of all $\mathcal{F}$-analytic sets is denoted by $\mathcal{A}(\mathcal{F})$. 7

It follows immediately from the definition that every $A \in \mathcal{A}(\mathcal{F})$ is contained in some element of $\mathcal{F}_\sigma$. In particular, the whole space F is $\mathcal{F}$-analytic if and only if it belongs to $\mathcal{F}_\sigma$ (8 below).

Definition 7 involves a variable compact space E. We show in the appendix that replacing E by the fixed compact space $\bar{\mathbb{N}}^{\mathbb{N}}$ ($\bar{\mathbb{N}}$ being the one point compactification of $\mathbb{N}$), or by $\bar{\mathbb{R}}$, leads to the same class of analytic sets. The same is true, on the other hand, if E is replaced by a variable semi-compact paved space $(E, \mathcal{E})$, as was done in the first edition. Finally, the $\mathcal{F}$-analytic sets are those which are constructed from Souslin's operation (A) applied to elements of $\mathcal{F}$.

---

(1) Bourbaki [2] (3rd edition), §6, no 4, Proposition 5. This proof was communicated to us by G. Mokobodzki.

8 THEOREM. $\mathcal{F} \subset \mathcal{A}(\mathcal{F})$; <u>the paving</u> $\mathcal{A}(\mathcal{F})$ <u>is closed under</u> $(\cup c, \cap c)$.

<u>Proof</u> : The first assertion is obvious. To establish the second, we consider a sequence $(A_n)_{n \in \mathbb{N}}$ of $\mathcal{F}$-analytic sets. There exist by definition, for each integer n :

- a compact metrizable space E_n, with its paving $\mathcal{K}(E_n) = \mathcal{E}_n$;
- a subset $B_n \in E_n \times F$, belonging to $(\mathcal{E}_n \times \mathcal{F})_{\sigma\delta}$ (and hence equal to the intersection of a sequence $(B_{nm})_{m \in \mathbb{N}}$ of elements of $(\mathcal{E}_n \times \mathcal{F})_\sigma$) whose projection onto F is A_n.

Let E be the compact space $\prod_n E_n$ with the paving $\mathcal{E} = \prod_n \mathcal{E}_n \subset \mathcal{K}(E)$. Let π be the projection of $E \times F$ onto F. We denote by C_n the cylinder of base B_n in $E \times F$, that is $(\prod_{n \neq m} E_m) \times B_n$ [(1)] ; then $\bigcap_n A_n = \pi(\bigcap_n C_n)$. The assertion concerning the operation $(\cap c)$ will therefore be established if we show that $\bigcap_n C_n$ belongs to $(\mathcal{E} \times \mathcal{F})_{\sigma\delta}$; which is obvious since every C_n belongs to $(\mathcal{E} \times \mathcal{F})_{\sigma\delta}$.

Now let E be the Alexandrov compactification of the topological sum $\sum_n E_n$, with the compact paving $\mathcal{E} = \sum_n \mathcal{E}_n \subset \mathcal{K}(E)$ and let π be the projection of $E \times F$ onto F. Then $\pi(\sum_n B_n) = \bigcup_n A_n$ (identifying $(\sum_n E_n) \times F$ to $\sum_n (E_n \times F)$). Hence it is sufficient to show that $\sum_n B_n \in (\mathcal{E} \times \mathcal{F})_{\sigma\delta}$. Now this set is equal to $\bigcap_m \sum_n B_{nm}$ and $\sum_n B_{nm}$ belongs to $(\mathcal{E} \times \mathcal{F})_\sigma$. Thus closure under $(\cup c)$ is established.

9 THEOREM. (a) <u>Let</u> $(E, \mathcal{E})$ <u>and</u> $(F, \mathcal{F})$ <u>be two paved sets</u> ; <u>we have</u>

$$\mathcal{A}(\mathcal{E}) \times \mathcal{A}(\mathcal{F}) \subset \mathcal{A}(\mathcal{E} \times \mathcal{F}).$$

(b) <u>Suppose that</u> E <u>is a compact metrizable space and</u> $\mathcal{E} = \mathcal{K}(E)$ <u>and let</u> A' <u>be an element of</u> $\mathcal{A}(\mathcal{E} \times \mathcal{F})$. <u>The projection</u> A <u>of</u> A' <u>onto</u> F <u>then is</u> $\mathcal{F}$-<u>analytic</u>.

<u>Proof</u> : Let A and B be two elements of $\mathcal{A}(\mathcal{E})$ and $\mathcal{A}(\mathcal{F})$ respectively ; and let A_1 and B_1 be two elements of $\mathcal{E}_\sigma$ and $\mathcal{F}_\sigma$ respectively, such that $A \subset A_1$, $B \subset B_1$. Then obviously $\mathcal{A}(\mathcal{E}) \times \mathcal{F} \subset \mathcal{A}(\mathcal{E} \times \mathcal{F})$, hence $\mathcal{A}(\mathcal{E}) \times \mathcal{F}_\sigma \subset \mathcal{A}(\mathcal{E} \times \mathcal{F})$ and similarly $\mathcal{E}_\sigma \times \mathcal{A}(\mathcal{F}) \subset \mathcal{A}(\mathcal{E} \times \mathcal{F})$ (8). Therefore

$$A \times B = (A \times B_1) \cap (A_1 \times B) \in \mathcal{A}(\mathcal{E} \times \mathcal{F})$$

and (a) is established.

We prove (b) : Since A' belongs to $\mathcal{A}(\mathcal{E} \times \mathcal{F})$, there exist an auxiliary compact metrizable space G (with its paving $\mathcal{K}(G) = \mathcal{G}$ and a subset A" of $G \times (E \times F)$ belonging to $(\mathcal{G} \times (\mathcal{E} \times \mathcal{F}))_{\sigma\delta}$, such that A' is the projection of A" onto $E \times F$. Then note that the paving $\mathcal{G} \times \mathcal{E}$ is contained in $\mathcal{K}(G \times E)$ and that A" is an element of $((\mathcal{G} \times \mathcal{E}) \times \mathcal{F})_{\sigma\delta}$ whose projection onto F is A.

10 THEOREM. <u>Let</u> $(F, \mathcal{F})$ <u>be a paved set and</u> $\mathcal{G}$ <u>be a paving such that</u> $\mathcal{F} \subset \mathcal{G} \subset \mathcal{A}(\mathcal{F})$. <u>Then</u> $\mathcal{A}(\mathcal{A}(\mathcal{F})) = \mathcal{A}(\mathcal{G}) = \mathcal{A}(\mathcal{F})$. <u>In particular</u>, $\mathcal{A}(\mathcal{G}) = \mathcal{A}(\mathcal{F})$ <u>if</u> $\mathcal{G}$ <u>is the closure of</u> $\mathcal{F}$ <u>under</u>

(1) We commit here an obvious abuse of notation ("commutativity" of a product of sets).

$(\cup c, \cap c)$.

Proof : Clearly $\mathcal{A}(\mathcal{A}(\mathcal{F})) \supset \mathcal{A}(\mathcal{G}) \supset \mathcal{A}(\mathcal{F})$. Let A be an $\mathcal{A}(\mathcal{F})$-analytic set ; there exist a compact metrizable space E (with its paving $\mathcal{K}(E) = \mathcal{E}$) and an element A' of $(\mathcal{E} \times \mathcal{A}(\mathcal{F}))_{\sigma\delta}$, such that A is the projection of A' onto F, Now $\mathcal{E} \times \mathcal{A}(\mathcal{F}) \subset \mathcal{A}(\mathcal{E}) \times \mathcal{A}(\mathcal{F}) \subset \mathcal{A}(\mathcal{E} \times \mathcal{F})$ (9) and hence $A' \in \mathcal{A}(\mathcal{E} \times \mathcal{F})$ (8). Applying 9 (b), we then see that A belongs to $\mathcal{A}(\mathcal{F})$.

More generally :

THEOREM. Let $(F,\mathcal{F})$ and $(G,\mathcal{G})$ be two paved sets and f be a mapping of F into G such that $f^{-1}(\mathcal{G}) \subset \mathcal{A}(\mathcal{F})$. Then $f^{-1}(\mathcal{A}(\mathcal{G})) \subset \mathcal{A}(\mathcal{F})$. 11

Proof : Let A be an element of $\mathcal{A}(\mathcal{G})$ and E a compact metrizable space (with its paving $\mathcal{K}(E) = \mathcal{E}$), such that there exists some $B \in (\mathcal{E} \times \mathcal{G})_{\sigma\delta}$ whose projection onto G is A. Let h denote the mapping $(x,y) \mapsto (x,f(y))$ of $E \times F$ into $E \times G$. The set $C = h^{-1}(B)$ obviously belongs to $(\mathcal{E} \times \mathcal{A}(\mathcal{F}))_{\sigma\delta} \subset (\mathcal{A}(\mathcal{E} \times \mathcal{F}))_{\sigma\delta} = \mathcal{A}(\mathcal{E} \times \mathcal{F})$ (9 and 8) ; since the set $f^{-1}(A)$ is the projection of C onto F, it is $\mathcal{F}$-analytic by 9.

THEOREM. $\mathcal{A}(\mathcal{F})$ contains the σ-field $\sigma(\mathcal{F})$ generated by $\mathcal{F}$ if and only if the complement of every element of $\mathcal{F}$ is $\mathcal{F}$-analytic. 12

Proof : The condition is obviously necessary. To show it is also sufficient, we consider the family $\mathcal{J}$ of all $\mathcal{F}$-analytic subsets of F whose complement is $\mathcal{F}$-analytic $\mathcal{J}$ is obviously a σ-field contained in $\mathcal{A}(\mathcal{F})$ and $\mathcal{F} \subset \mathcal{J}$. Hence $\sigma(\mathcal{F}) \subset \mathcal{J} \subset \mathcal{A}(\mathcal{F})$.

As an easy corollary to Theorems 9 and 12, we can show the analyticity of some projection sets which occur frequently in probability theory. We denote by $\mathcal{B}$ the Borel σ-field of $\mathbb{R}$ and by $\mathcal{K}$ the paving consisting of all compact subsets of $\mathbb{R}$; $\mathbb{R}$ can in fact be replaced by any LCC space (in particular by any metrizable compact space).

THEOREM. (1) $\mathcal{B} \subset \mathcal{A}(\mathcal{K})$, $\mathcal{A}(\mathcal{B}) = \mathcal{A}(\mathcal{K})$. 13

(2) Let $(\Omega,\mathcal{F})$ be a measurable space. The product σ-field $\mathcal{G} = \mathcal{B} \times \mathcal{F}$ on $\mathbb{R} \times \Omega$ is contained in $\mathcal{A}(\mathcal{K} \times \mathcal{F})$.

(3) The projection onto Ω of an element of $\mathcal{G}$ (or, more generally, of $\mathcal{A}(\mathcal{G})$) is $\mathcal{F}$-analytic.

Proof : We note that the complement of a compact subset of $\mathbb{R}$ is a countable union of compact sets and that $\mathcal{B}$ is the σ-field generated by $\mathcal{K}$; by 12, we have $\mathcal{B} \subset \mathcal{A}(\mathcal{K})$, hence $\mathcal{A}(\mathcal{B}) \subset \mathcal{A}(\mathcal{K})$ by 10 and finally $\mathcal{A}(\mathcal{B}) = \mathcal{A}(\mathcal{K})$. Similarly, the product paving $\mathcal{K} \times \mathcal{F}$ generates the σ-field $\mathcal{G}$ and the complement of an element of $\mathcal{K} \times \mathcal{F}$ is a countable union of elements of $\mathcal{K} \times \mathcal{F}$. Hence $\mathcal{G} \subset \mathcal{A}(\mathcal{K} \times \mathcal{F})$ (and in fact, as above, $\mathcal{A}(\mathcal{G}) = \mathcal{A}(\mathcal{K} \times \mathcal{F})$). Finally, (3) follows from 9.

We now have everything we need to prove and apply Choquet's theorem on capacities. But we do not end this paragraph here. We first present the separation theorem for analytic sets, which is the last one of our "abstract" results. We then pass to the study of analytic sets in special spaces, where the separation theorem leads to simple and very useful results (the Souslin-Lusin Theorem, Blackwell's Theorem).

The separation theorem

Here the reader has a choice between two proofs : the following one, which is "elementary" and a little clumsy, and another one in the next section (43) which shows that the separation theorem is a particular case of capacitability (an idea of Sion presented in a simplified way). The substance of the two proofs is however the same.

14 We denote by $(F,\mathcal{F})$ a paved set and by $\mathcal{C}(\mathcal{F})$ the closure of $\mathcal{F}$ under $(\cup c, \cap c)$. For example, if F is a LCC space and $\mathcal{F}$ is the paving of compact subsets of F, $\mathcal{C}(\mathcal{F})$ is the Borel σ-field.

Two subsets A and A' of F are called <u>separable by elements of</u> $\mathcal{C}(\mathcal{F})$ if there exist two disjoint elements of $\mathcal{C}(\mathcal{F})$ containing respectively A and A'.

THEOREM. *Suppose that the paving $\mathcal{F}$ is semi-compact and let A and A' be two disjoint $\mathcal{F}$-analytic sets. Then A and A' can be separated by elements of $\mathcal{C}(\mathcal{F})$.*

Proof : It is convenient for the proof to assume that $F \in \mathcal{F}$. A moment's reflection will show that this further condition does not weaken the statement.

We begin with an auxiliary result : *let (C_n) and (D_n) be two sequences of subsets of F such that C_n and D_m are separable by elements of $\mathcal{C}(\mathcal{F})$ for every pair (n,m); then the sets $\bigcup_n C_n$ and $\bigcup_m D_m$ are separable.*

We choose indeed, for each pair (n,m), two elements E_{nm}, F_{nm} of $\mathcal{C}(\mathcal{F})$ such that $C_n \subset E_{nm}$, $D_m \subset F_{nm}$, $E_{nm} \cap F_{nm} = \emptyset$. We set

$$E' = \bigcup_n \bigcap_m E_{nm} \; , \; F' = \bigcup_p \bigcap_n F_{np} \; ;$$

these sets belong to $\mathcal{C}(\mathcal{F})$ and $\bigcup_n C_n \subset E'$, $\bigcup_m D_m \subset F'$ and $E' \cap F' = \emptyset$.

Having established this lemma, we consider two disjoint $\mathcal{F}$-analytic sets A and A'. By a preliminary construction (that of a product) we can assume that there exists one single compact metrizable space E, with its paving $\mathcal{K}(E) = \mathcal{E}$, such that A and A' are respectively the projections of sets :

$$J = \bigcap_n \bigcup_m J_{nm} \; , \; J' = \bigcap_n \bigcup_m J'_{nm}$$

where the sets $J_{nm} = E_{nm} \times F_{nm}$, $J'_{nm} = E'_{nm} \times F'_{nm}$ belong to $\mathcal{E} \times \mathcal{F}$. To abbreviate, let us say that two subsets of $E \times F$ are <u>separable</u> if their projections onto F are separable by elements of $\mathcal{C}(\mathcal{F})$. We then assume that J and J' are not separable and deduce that A and A' are not disjoint, contrary to the hypothesis.

Let $m_1, m_2, \ldots, m_i$ be integers. We set

$$L_{m_1 m_2 \ldots m_i} = J_{1m_1} \cap J_{2m_2} \cap \ldots \cap J_{im_i} \cap \Big(\bigcap_{n>i} \bigcup_m J_{nm}\Big).$$

$L'_{m'_1m'_2\ldots m'_i}$ is defined similarly. Since $J = \bigcup_{m_1} L_{m_1}$, $J' = \bigcup_{m'_1} L'_{m'_1}$ and J and J' are not separable, the above lemma implies the existence of two integers m_1, m'_1 such that L_{m_1} and $L'_{m'_1}$ are not separable ; but $L_{m_1} = \bigcup_{m_i} L_{m_1m_i}$, $L'_{m'_1} = \bigcup_{m'_i} L'_{m'_1m'_i}$. Hence there exist two integers m_2 and m'_2 such that $L_{m_1m_2}$ and $L'_{m'_1m'_2}$ are not separable. Thus we construct inductively two infinite sequences $m_1, m_2, \ldots, m'_1, m'_2 \ldots$ such that $L_{m_1m_2\ldots m_i}$ and $L'_{m'_1m'_2\ldots m'_i}$ aren't separable.

These sets cannot be empty, since every subset of $E \times F$ can be separated from the empty set (here we use the assumption that $F \in \mathcal{C}(\mathcal{F})$). Hence

$$E_{1m_1} \cap E_{2m_2} \cap \ldots \cap E_{im_i} \neq \emptyset, \quad E'_{1m'_1} \cap E'_{2m'_2} \cap \ldots \cap E'_{im'_i} \neq \emptyset.$$

Similarly,

$$(F_{1m_1} \cap F_{2m_2} \cap \ldots \cap F_{im_i}) \cap (F'_{1m'_1} \cap F'_{2m'_2} \cap \ldots \cap F'_{im'_i}) \neq \emptyset$$

because the two sets in brackets belong to $\mathcal{C}(\mathcal{F})$ and $L_{m_1m_2\ldots m_i}$ and $L'_{m'_1m'_2\ldots m'_i}$ are not separable. The pavings $\mathcal{E}$ and $\mathcal{F}$ are semi-compact, so there exist $x \in \bigcap_i E_{im_i}$, $x' \in \bigcap_i E'_{im'_i}$ and $y \in \bigcap_i (F_{im_i} \cap F'_{im'_i})$. Then $(x,y) \in J$, $(x',y) \in J'$ and finally $y \in A \cap A'$, the desired contradiction.

Souslin measurable spaces, etc.

Until now we have been concerned with $\mathcal{F}$-analytic subsets of a paved set $(F,\mathcal{F})$: a 15
statement of the type "A is $\mathcal{F}$-analytic" does not express an intrinsic property of the set A but rather something about its position within a larger set. We are going to change slightly our point of view: $(F,\mathcal{F})$ will be a <u>measurable space</u> and we shall study subsets A of F characterized by intrinsic properties of the measurable space $(A,\mathcal{F}|_A)$. In paragraph 3 we shall complete this with considerations of a topological nature.

We are grateful to Yen Kia An for his comments on this part of our text : finally, it was a project of his that led to the form adopted below.

We first recall a few facts from chapter I. Two measurable spaces $(E,\mathcal{E})$, $(F,\mathcal{F})$ are said to be <u>isomorphic</u> if there exists a bijection between E and F which is measurable and has a measurable inverse. A space $(F,\mathcal{F})$ is called <u>Hausdorff</u> if the atoms of $\mathcal{F}$ are the points of F. A Hausdorff separable measurable space $(F,\mathcal{F})$ is isomorphic to a space $(U,\mathcal{B}(U))$, where U is a (not necessarily Borel) subset of $\mathbb{R}$ (I.11).

We omit the proof of the following lemma, which is immediate from the definition of analytic sets.

LEMMA. <u>Let</u> $(F,\mathcal{F})$ <u>be a paved set</u>, E <u>be a subset of</u> F <u>and</u> $\mathcal{E}$ <u>be the paving</u> $\mathcal{F}|\mathcal{E}$, <u>i.e. the trace of</u> $\mathcal{F}$ <u>on</u> E. <u>Then</u> $\mathcal{A}(\mathcal{E}) = \mathcal{A}(\mathcal{F})|_E$.

Given a topological space E, we denote by $\mathcal{K}(E)$, $\mathcal{G}(E)$, $\mathcal{B}(E)$ - or simply $\mathcal{K},\mathcal{G},\mathcal{B}$ - the pavings consisting of the compact, the open, the Borel subsets of E. If E is metrizable, the complement of an open set is a $\mathcal{G}_\delta$, hence $\mathcal{G} \subset \mathcal{B} \subset \mathcal{A}(\mathcal{G})$ (12) and $\mathcal{A}(\mathcal{B}) = \mathcal{A}(\mathcal{G})$ (10). The latter paving will be denoted by $\mathcal{A}(E)$ or $\mathcal{A}$ and its elements will be called analytic in E. In the metrizable case (1), these are the same as the $\mathcal{f}$-analytic sets, where $\mathcal{f}$ is the paving of closed subsets of E, and the same as the $\mathcal{K}$-analytic sets if E is compact (or LCC).

16 DEFINITION. (a) A metrizable topological space is said to be Lusin (resp. Souslin, cosouslin) if it is homeomorphic to a Borel subset (resp. an analytic subset a complement of an analytic subset)(2) of a compact metrizable space.

(b) A measurable space $(F,\mathcal{F})$ is said to be Lusin (resp. Souslin, cosouslin) if it is isomorphic to a measurable space $(H,\mathcal{B}(H))$, where H is a Lusin (resp. Souslin, cosouslin) metrizable space.

(c) In a Hausdorff measurable space $(F,\mathcal{F})$, a set E is said to be Lusin (resp. Souslin, cosouslin) if the measurable space $(E,\mathcal{F}|_E)$ is Lusin (resp. Souslin, cosouslin). We denote by $\mathcal{L}(\mathcal{F})$, $\mathcal{S}(\mathcal{F})$, $\mathcal{S}'(\mathcal{F})$ the pavings consisting of the Lusin, Souslin, cosouslin sets in $(F,\mathcal{F})$.

A Lusin metrizable (resp. measurable) space is both Souslin and cosouslin (12) and we shall see later that the converse is true. Every Lusin, Souslin or cosouslin metrizable (resp. measurable) space is separable and Hausdorff.

The introduction of cosouslin spaces is not a mere luxury. We shall see in chapter IV that the usual spaces of paths are either Lusin or cosouslin.

We give in no. 67 a more general definition of Lusin (resp. Souslin) topological spaces : these spaces still are Hausdorff but, not necessarily metrizable ; however, their Borel σ-field is a Lusin (resp. Souslin) σ-field in the sense of (b).

We first give a means of constructing Lusin measurable spaces, etc... (which will be made more precise in no. 19) and the most usual example of a non-compact Lusin metrizable space.

17 THEOREM. Let $(F,\mathcal{F})$ be a measurable space. If F is Lusin, then $\mathcal{F} \subset \mathcal{L}(\mathcal{F})$. If F is Souslin, then $\mathcal{A}(\mathcal{F}) \subset \mathcal{S}(\mathcal{F})$. If F is cosouslin, the complement of every element of $\mathcal{A}(\mathcal{F})$ belongs to $\mathcal{S}'(\mathcal{F})$.

Every Polish space is Lusin (and hence so is every Borel subspace of a Polish space).

Proof : We can assume that F is a subset of a compact metrizable space C, with $\mathcal{F}$ the trace of $\mathcal{B}(C)$ on F. Then the Lusin case is obvious and the two other cases reduce to the Lemma in no. 15 : $\mathcal{A}(\mathcal{F})$ is the trace of $\mathcal{A}(C)$ on F.

For the second part of the statement, we just copy Bourbaki. Let P be a complete separable metric space ; we imbed P as a topological (not a metric!) subspace of a

(1) In the non-metrizable case, there are at least 5 possible definitions of analytic sets and we prefer to say no more.

(2) "complement of analytic" is often abbreviated into "coanalytic".

compact metrizable space C, for example of the cube $[0,1]^{\mathbb{N}}$, and denote by U_n the set of all $x \in \bar{P}$ possessing some neighbourhood in C whose trace on P has a diameter $< 1/n$ relative to the metric of P. U_n is open in $\bar{P}$ and hence a $\mathcal{G}_\delta$ in C ; since P is complete it is equal to $\bigcap_n U_n$ and P is therefore a $\mathcal{G}_\delta$ in C.

Conversely, it can be shown that any $\mathcal{G}_\delta$ of a compact metrizable space is Polish (Bourbaki [3], General Topology, Chapter IX, §6, no. 1, Theorem 1).

The following theorem on direct images - which can be compared with 9 - is a useful 18
result, but less deep than Theorems 21 and 26. Note that its proof does not use the separation theorem. Parts (a) and (b) are technical lemmas and the important part of the statement is (c) and (d).

THEOREM. Let $(F,\mathcal{F})$ and $(F',\mathcal{F}')$ be two Hausdorff separable measurable spaces. Assume (1) that F and F' are imbedded in compact metric spaces C and C' and that $\mathcal{F} = \mathcal{B}(C)|_F$, $\mathcal{F}' = \mathcal{B}(C')|_{F'}$. Let f be a measurable mapping of F into F'. Then

(a) f can be extended to a Borel mapping g of C into C' ;

(b) If further f is an isomorphism between F and F', there exist two Borel subsets $B \supset F$, $B' \supset F'$ of C and C' respectively such that g induces an isomorphism between B and B' ;

(c) for all $A \in \mathcal{S}(\mathcal{F})$, we have $f(A) \in \mathcal{S}(\mathcal{F}')$;

(d) $\mathcal{S}(\mathcal{F}) \subset \mathcal{A}(F)$, and $\mathcal{S}(\mathcal{F}) = \mathcal{A}(F)$ if and only if F is Souslin. In particular, if F is Souslin, then $f(\mathcal{A}(\mathcal{F})) \subset \mathcal{A}(\mathcal{F}')$.

Proof : (a) Let i be the canonical injection of F into C. Then $\mathcal{S}(F) = \sigma(i)$ and hence f, considered as a mapping of F into C', is $\sigma(i)$-measurable. By I.18, there exists a Borel mapping g of C into C' such that $f = g \circ i$ and g is the required extension.

(b) Similarly, if f is a Borel isomorphism, the inverse mapping $f' = f^{-1}$ of F' into F can be extended to a Borel mapping g' of C' into C. It suffices to take $B = \{x \in C : g'(g(x)) = x\}$, $B' = \{x' \in C' : g(g'(x')) = x'\}$ (I.12) and the required isomorphism is $g|_B$ with inverse isomorphism $g'|_{B'}$.

(c) We return to the situation in (a). The graph G of g belongs to the product σ-field $\mathcal{B}(C) \times \mathcal{B}(C')$ (I.12) and is therefore $\mathcal{K}(C) \times \mathcal{K}(C')$-analytic (13). If $(F,\mathcal{F})$ is itself Souslin, we choose C and the imbedding such that $F \in \mathcal{A}(C) = \mathcal{A}(\mathcal{K}(C))$ - this is possible by the very definition of Souslin measurable spaces. Then $(F \times C') \cap G$ is $\mathcal{K}(C) \times \mathcal{K}(C')$-analytic (9(a)). By projecting onto C', we see that $g(F) = f(F)$ is analytic in C' (9(b)). Since C' is a compact metric space, f(F) is a Souslin measurable space contained in F' and hence $f(F) \in \mathcal{S}(\mathcal{F}')$. On the other hand, f(F) is $\mathcal{A}(C')$-analytic and contained in F' and hence belongs to $\mathcal{A}(\mathcal{F}')$ (no. 15).

If now $A \in \mathcal{S}(\mathcal{F})$, this result applied to $f|_A$ shows that f(A) belongs to $\mathcal{S}(\mathcal{F}') \cap \mathcal{A}(F')$.

(1) This implies no restriction on generality : F and F' can always be so imbedded in the interval [0,1] (I.11)

(d) To keep the same notation, we show that $\mathcal{S}(\mathcal{F}') \subset \mathcal{A}(F')$. Let $B \in \mathcal{S}(\mathcal{F}')$. By definition of Souslin spaces, there exist a compact metrizable space C, a subset F analytic in C and a Borel isomorphism f of F onto B. But then we have seen that B belongs to $\mathcal{S}(\mathcal{F}') \cap \mathcal{A}(F')$ and hence $B \in \mathcal{A}(F')$. If F' is Souslin, we have seen in no. 17 that $\mathcal{A}(F') \subset \mathcal{S}(\mathcal{F}')$, whence the equality. Conversely, if $\mathcal{S}(\mathcal{F}') = \mathcal{A}(F')$, as $F' \in \mathcal{A}(F')$, F' is Souslin.

The following theorem restates part (d) of 18 and gives analogous results for $\mathcal{L}(\mathcal{E})$ and $\mathcal{S}'(\mathcal{E})$. Its importance must be emphasized. As said in 15, the fact that a subset A belongs to $\mathcal{L}(\mathcal{E})$, for example, is an intrinsic property of the measurable space $(A,\mathcal{A})$, where $\mathcal{A}$ is the trace σ-field of $\mathcal{E}$ on A. The fact that A belongs to $\mathcal{E}$, on the other hand, is a position property of the set A in E, Hence inclusion (1) expresses the fact that in whatever way the measurable space $(A,\mathcal{A})$ is imbedded in a larger space $(E,\mathcal{E})$, A belongs to the σ-field $\mathcal{E}$. Similarily for (2) and (3) [1].

For conciseness of notation, we denote here by $\mathcal{A}'(\mathcal{E})$ the family of subsets of E whose complements are $\mathcal{E}$-analytic.

19 THEOREM. Let $(E,\mathcal{E})$ be a Hausdorff separable measurable space. Then

(1) $\mathcal{L}(\mathcal{E}) \subset \mathcal{E}$, with equality if and only if E is Lusin ;

(2) $\mathcal{S}(\mathcal{E}) \subset \mathcal{A}(\mathcal{E})$, with equality if and only if E is Souslin ;

(3) $\mathcal{S}'(\mathcal{E}) \subset \mathcal{A}'(\mathcal{E})$, with equality if and only if E is cosouslin.

Proof : We go back to Theorem 18 with F' = E. By definition, if $A \subset E$ belongs to $\mathcal{L}(\mathcal{E})$ (resp. $\mathcal{S}(\mathcal{E})$, $\mathcal{S}'(\mathcal{E})$), there exist a compact metrizable space C, a Borel subset F of C (resp. a subset F analytic in C, coanalytic in C) and a Borel isomorphism f of F onto E = F'. By 18 (b), f extends to an isomorphism g between two Borel subsets B and B' of C and C'. Then g(F) = F' is (by isomorphism) Borel in B' (resp. analytic, coanalytic) and is contained in E = F' and hence belongs to $\mathcal{E}(\mathcal{A}(\mathcal{E}), \mathcal{A}'(\mathcal{E}))$. The remainder is obvious.

20 REMARKS. (a) Using I.11, we may identify the measurable space $(E,\mathcal{E})$ to a subset of the interval [0,1]) still denoted by E - with its Borel σ-field. The elements of $\mathcal{L}(\mathcal{E})$, $\mathcal{S}(\mathcal{E})$, $\mathcal{S}'(\mathcal{E})$ are then interpreted, by 19, as the Borel sets, analytic sets, coanalytic sets of [0,1], contained in E. It follows immediately that these pavings on E are closed, under $(\cup c, \cap c)$ and moreover that if E is both Souslin and cosouslin, it is both analytic and coanalytic in [0,1], hence Borel in [0,1] by the separation theorem, and finally Lusin.

(b) In view of future references, we give explicitly an easy consequence of 19 ; we do not write down the "cosouslin" statement.

THEOREM. (1) Every Lusin (Souslin) measurable space is isomorphic to an (analytic) Borel subset of [0,1].

(2) Every Lusin (Souslin) metrizable space is homeomorphic to an (analytic)

(1) Compare this with the result, topological in nature, of no. 17 : however a Polish space P is imbedded in a compact metrizable space K, P is a $\mathcal{G}_\delta$ in K.

Borel subspace of the cube $[0,1]^{\mathbb{N}}$.

Proof : For (1), use I.11 and 19. For (2), imbed the space into the cube and use 19.

In fact any uncountable Lusin measurable space is isomorphic to the whole of $[0,1]$ (see the appendix).

The Souslin-Lusin theorem

To apply 19, it is necessary to know whether a given subset A of E belongs for example to $\mathcal{L}(\mathcal{E})$: this means that we are able to construct between A and some known Lusin space L a measurable bijection f with a measurable inverse. The Souslin-Lusin theorem will spare us worrying about f^{-1}. It is one of the great steam-hammers of measure theory and we shall use it whenever possible (even for cracking nuts). (21)

THEOREM. Let $(F,\mathcal{F})$ and $(F',\mathcal{F}')$ be two Hausdorff separable measurable spaces and h be an injective measurable mapping of F into F'.

(1) If F is Souslin, h is an isomorphism of F onto h(F).

(2) Further, if F is Lusin, then $h(\mathcal{F}) \subset \mathcal{L}(\mathcal{F}') \subset \mathcal{F}'$.

Proof : By I.11, there is no loss in generality in supposing that $(F',\mathcal{F}')$ is the interval $[0,1]$ with its Borel σ-field. Let $A \in \mathcal{F}$. The sets h(A) and $h(A^c)$ are Souslin (18) and disjoint since h is injective. By the separation theorem 14, since the Souslin subsets of $[0,1]$ are $\mathcal{K}$-analytic, h(A) and $h(A^c)$ can be separated by two Borel subsets B and B' of $[0,1]$. Then $h^{-1}(B)$ and $h^{-1}(B')$ are elements of $\mathcal{F}$ which separate A and A^c and hence $A = h^{-1}(B)$, $A^c = h^{-1}(B')$ and $h(A) = B \cap h(F)$. Thus $h(A) \in \mathcal{F}'$, which means that h is a measurable isomorphism. The rest of the statement follows from 19. (1).

How far can the injectivity hypothesis on f be weakened? We state without proof (cf. Hausdorff, [1]) an extension of 21, due to Lusin, which is much more difficult.

THEOREM. With the hypotheses of 21, suppose that F is Lusin and that, for all $y \in F'$, 21a
$h^{-1}(\{y\})$ is at most countable. Then the direct image under h of every element of $\mathcal{F}$ is measurable in F' and there exists a countable partition (F_n) of F into measurable sets, on each of which h is injective.

On the other hand, if we do not demand that the image under h of every element of $\mathcal{F}$ be measurable, but only that h(F) be, we have the following results due to Novikov and Kunugui (resp. Arsenin, Čegolkov and Čoban) which we state in topological language.

THEOREM. Let F and F' be two separable metrizable spaces and h a Borel mapping of F 21b
into F'. If F is Lusin and, for all $y \in F'$, $h^{-1}(\{y\})$ is compact in F (resp. more generally, is a countable union of compact subsets of F), the image h(F) is Borel in F'. Further, there then exists a Borel subset B of F such that the restriction of h to B is injective with image equal to h(F).

in 21b, compact sets can even be replaced by $\mathcal{K}_\sigma$ sets. See Saint-Raymond, Bull. Soc. Math. France, 104, 1976, p. 389-400.

Here is a corollary to 21, particularly useful under the from of the remark following the proof. We use topological language.

22 THEOREM. Let Ω and E be two separable metrizable spaces and let f be a mapping of Ω into E. If the graph of f is a Souslin subspace of $\Omega \times E$, then Ω is Souslin and f is a Borel mapping of Ω into E.

Proof : Let G be the graph of f and π the projection of $\Omega \times E$ onto Ω. By 18.(c), the space Ω, which is equal to $\pi(G)$, is Souslin. Moreover, since the restriction of π to G is an injective Borel mapping of G onto Ω, the measurable spaces G and Ω are isomorphic by 21. Then, if B is a Borel subset of E, $f^{-1}(B)$ is equal to the projection onto Ω of the Borel subset $G \cap (\Omega \times B)$ of G and hence is Borel in E. Thus f is Borel.

REMARK. Suppose that Ω and E are Souslin : then they can be imbedded in compact metrizable spaces K and L as analytic subspaces. Then $\Omega \times E$ is analytic in $K \times L$ (9.(a)) and the measurable space $(\Omega \times E, \mathcal{B}(\Omega \times E))$ is Souslin. Every element of $\mathcal{B}(\Omega \times E) \subset \mathcal{A}(\mathcal{B}(\Omega \times E)$ then is Souslin (18.(d)). Thus

if Ω and E are Souslin and the graph of f is a Borel subset of $\Omega \times E$, f is Borel. This is the converse of I.12(b).

We now return to the separation theorem, which was the essential tool for the proof of theorem 21. The following very simple modification frees it from the "compactness" hypothesis of no. 14. We shall not need it later, however, so it can be omitted without harm. See also the remark in no. 68.

23 THEOREM. Let $(F,\mathcal{F})$ be a separable Hausdorff measurable space. Let A and A' be two disjoint elements of $\mathcal{S}(\mathcal{F})$. Then there exist two disjoint elements B and B' of $\mathcal{F}$ such that $A \subset B$, $A' \subset B'$.

Proof : We imbed F in a compact metrizable space C. By 18, A and A' are $\mathcal{K}(C)$-analytic. By 14, they are separable by Borel subsets of C, whose traces on F are the required sets.

Blackwell spaces

For elementary definitions (atoms, associated Hausdorff space,...) used in this section, see I.9-10.

Blackwell spaces were introducted by Blackwell in [1], under the name of Lusin spaces, which led to confusion (the term Blackwell space is itself used by some authors with a slightly different meaning, but still our terminology conforms to general use).

24 DEFINITION. A measurable space $(\Omega,\mathcal{F})$ is a Blackwell space if the associated Hausdorff space $(\dot{\Omega},\dot{\mathcal{F}})$ is Souslin.

25 THEOREM. $(\Omega,\mathcal{F})$ is a Blackwell space, if and only if $\mathcal{F}$ is separable and the following property holds.

(25.1) *For every measurable mapping* f *of* $(\Omega,\mathcal{F})$ *into* R *and every* $A \in \mathcal{F}$, *the image* f(A) *is analytic in* $\mathbb{R}$.

Proof : We can reduce to the case where $(\Omega,\mathcal{F})$ is Hausdorff. Then saying that $(\Omega,\mathcal{F})$ is a Blackwell space amounts to saying that it is Souslin.

First, if $(\Omega,\mathcal{F})$ is Souslin, we know that $\mathcal{F}$ is separable and that (25.1) is satisfied (18). Conversely, if $\mathcal{F}$ is separable and Hausdorff, there exists an isomorphism f of $(\Omega,\mathcal{F})$ into a subset of $\mathbb{R}$ (I.11), which is analytic by (25.1), and $(\Omega,\mathcal{F})$ is Souslin.

The following result, which we shall henceforth call *Blackwell's Theorem*, is quite useful : it reduces inclusion between Blackwell σ-fields to comparing their atoms.

THEOREM. *Let* $(\Omega,\mathcal{F})$ *be a Blackwell space*, $\mathcal{G}$ *a sub-*σ*-field of* $\mathcal{F}$ *and* $\mathcal{S}$ *a separable sub-* (26)
σ*-field of* $\mathcal{F}$. *Then* $\mathcal{G} \subset \mathcal{S}$, *if and only if every atom of* $\mathcal{G}$ *is a union of atoms of* $\mathcal{S}$.

In particular, a $\mathcal{F}$*-measurable real function* g *is* $\mathcal{S}$*-measurable, if and only if* g *is constant on every atom of* $\mathcal{S}$.

Proof : It suffices to establish the first assertion ; the second follows by taking $\mathcal{G} = \sigma(g)$.

The condition is obviously necessary. To see that it is sufficient, we take an arbitrary element B of $\mathcal{G}$ and denote by $\mathcal{J}$ the σ-field generated by $\mathcal{S}$ and B ; $\mathcal{J}$ is separable and satisfies (25.1) as a sub-σ-field of $\mathcal{F}$ and hence $(\Omega,\mathcal{F})$ is a Blackwell space. On the other hand, the two σ-fields $\mathcal{S}$ and $\mathcal{J}$ determine the same equivalence relation on Ω. Passing to the quotient, we are reduced to showing that the identity mapping - which is obviously measurable - of the Souslin space $(\dot{\Omega},\dot{\mathcal{F}})$ onto the Souslin space $(\dot{\Omega},\dot{\mathcal{S}})$ is a Borel isomorphism, and that was seen in 22.

REMARKS. (a) The theorem is no longer true if $\mathcal{S}$ isn't assumed to be separable. For example, let $\Omega = \mathbb{R}$, $\mathcal{F} = \mathcal{G} = \mathcal{B}(\mathbb{R})$ and $\mathcal{S}$ be the sub -σ-field of $\mathcal{F}$ generated by the sets $\{x\}$, $x \in \mathbb{R}$; the two σ-fields $\mathcal{S}$ and $\mathcal{F}$ have the same atoms but are not equal.

(b) Let f and g be two real random variables on a Blackwell space $(\Omega,\mathcal{F})$ and assume there exists a functional relation $g = h \circ f$ between f and g, where h is any mapping of $\mathbb{R}$ into $\mathbb{R}$; there then exists a functional relation $g = h' \circ f$, where h' is Borel. The σ-field $\sigma(f)$ is indeed separable and g is constant on every atom of $\sigma(f)$; hence g is $\sigma(f)$-measurable and we may apply 18.

2. CAPACITIES

The section can be understood whith little of the preceding material : we only need the definition of analytic sets and some elementary properties of compact pavings (6).

DEFINITION. *Let* F *be a set with a paving* $\mathcal{F}$ *closed under* $(\cup f,\cap f)$. *A* (Choquet) capacity 27
on F ($\mathcal{F}$*-capacity if there is any ambiguity) is an extended real valued set function*

I, defined for all subsets of F, with the following properties :

(a) I is increasing ($A \subset B \Rightarrow I(A) \leq I(B)$) ;

(b) for every increasing sequence $(A_n)_{n \in \mathbb{N}}$ of subsets of F,

$$I(\bigcup_n A_n) = \sup_n I(A_n) \ ; \tag{27.1}$$

(c) for every decreasing sequence $(A_n)_{n \in \mathbb{N}}$ of elements of $\mathfrak{F}$,

$$I(\bigcap_n A_n) = \inf_n I(A_n). \tag{27.2}$$

A subset A of F is called capacitable if

$$I(A) = \sup_{\substack{B \in \mathfrak{F}_\delta \\ B \subset A}} I(B). \tag{27.3}$$

In classical potential theory the "Newtonian outer capacity" is a Choquet capacity relative to the paving $\mathfrak{F}$ of all compact subsets of $\mathbb{R}^n$ ($n \geq 3$). Then $\mathfrak{F}_\delta = \mathfrak{F}$ and equation (27.3) means that the outer capacity of A can also be estimated "from inside". For details, see Brelot [1] or [2].

(28) THEOREM. (Choquet). *Let I be an $\mathfrak{F}$-capacity. Every $\mathfrak{F}$-analytic set is capacitable relative to I.*

We shall follow Bourbaki [(1)] in the proof of this theorem, which involves two lemmas.

LEMMA 1. *Every element of $\mathfrak{F}_{\sigma\delta}$ is capacitable relative to I.*

Proof : Let A be an element of $\mathfrak{F}_{\sigma\delta}$ such that $I(A) > -\infty$ [(2)] ; A is the intersection of a sequence $(A_n)_{n \geq 1}$ of elements of $\mathfrak{F}_\sigma$ and each A_n is the union of an *increasing* sequence $(A_{nm})_{m \geq 1}$ of elements of $\mathfrak{F}$. We show that there exists, for every number $a < I(A)$, an element B of $\mathfrak{F}_\delta$ such that $B \subset A$, $I(B) \geq a$. We first prove the existence of a sequence $(B_n)_{n \geq 1}$ of elements of $\mathfrak{F}$ such that $B_n \subset A_n$ and $I(C_n) > a$, ,where $C_n = A \cap B_1 \cap B_2 \cap \ldots \cap B_n$.

To construct B_1, we have by (27.1)

$$I(A) = I(A \cap A_1) = \sup_m I(A \cap A_m).$$

We then take B_1 to be one of the sets A_{1m}, where m is chosen sufficiently large so that $I(A \cap A_{1m}) > a$.

We then suppose that the construction has been made up to the (n-1) th term We have by hypothesis $C_{n-1} \subset A$, $I(C_{n-1}) > a$. Consequently :

$$I(C_{n-1}) = I(C_{n-1} \cap A_n) = \sup_m I(C_{n-1} \cap A_{nm}).$$

Then we take B_n to be one of the sets A_{nm}, where m is sufficiently large, so that $I(C_{n-1} \cap A_{nm}) = I(C_n) > a$.

---

(1) [3], §.6, Proposition 14

(2) The statement is trivial if $I(A) = -\infty$, for then $I(\emptyset) = -\infty$ and $\emptyset$ belongs to $\mathfrak{F}$.

Having constructed the sequence $(B_n)_{n \geq 1}$, we set $B'_n = B_1 \cap B_2 \cap \ldots \cap B_n$ and

$$B = \bigcap_n B_n = \bigcap_n B'_n.$$

The sets B'_n belong to $\mathcal{F}$ and decrease and we have $C_n \subset B'_n$: hence $I(B'_n) > a$ and $I(B) \geq a$ by (27.2). We have $B_n \subset A_n$ and hence $B \subset A$. Finally the set B satisfies the required conditions and the lemma is established.

Now let A be $\mathcal{F}$-analytic. There exist a compact metric space E with its compact paving $\mathcal{K}(E) = \mathcal{E}$ and an element B of $(\mathcal{E} \times \mathcal{F})_{\sigma\delta}$ such that the projection of B onto F is equal to A. Let π denote the projection of $E \times F$ onto F and $\mathcal{H}$ denote the paving consisting of all finite unions of elements of $\mathcal{E} \times \mathcal{F}$. By 4, there is no loss of generality in supposing that $\mathcal{E}$ is closed under $(\cup f, \cap f)$ and then $\mathcal{H}$ is closed under $(\cup f, \cap f)$.

LEMMA 2. <u>The set function</u> J <u>defined for all</u> $H \subset E \times F$ <u>by</u> :

$$J(H) = I(\pi(H))$$

<u>is an</u> $\mathcal{H}$<u>-capacity on</u> $E \times F$.

<u>Proof</u> : The function J is obviously increasing and satisfies (27.1). Property (27.2) follows immediately from the relation :

$$\bigcap_n \pi(B_n) = \pi(\bigcap_n B_n)$$

which holds, according to 6, for every decreasing sequence $(B_n)_{n \in \mathbb{N}}$ of elements of $\mathcal{H}$.

We can now complete the proof. Since B is capacitable relative to J by Lemma 1, there exists an element D of $\mathcal{H}_\delta$ such that $D \subset B$, $J(D) \geq J(B) - \varepsilon$ $(\varepsilon > 0)$. Let C be the set $\pi(D)$: the above equality shows that C is an element of $\mathcal{F}_\delta$ and we have $C \subset A$, $I(C) \geq I(A) - \varepsilon$.

It is interesting to analyze the above proof following Sion [1]. Let $\mathcal{C}$ be the class 29
of all sets A such that $I(A) > a$: $\mathcal{C}$ has the properties :

(29.1) $A \in \mathcal{C}$, $A \subset B \Rightarrow B \in \mathcal{C}$

(29.2) if (A_n) is an increasing sequence of subsets of F, whose union belongs to $\mathcal{C}$, then some A_n belongs to $\mathcal{C}$.

On the other hand, the property we established can be stated as follows :

(29.3) if an $\mathcal{F}$-analytic set belongs to $\mathcal{C}$, it contains the intersection of a decreasing sequence of elements of $\mathcal{F} \cap \mathcal{C}$.

The proof rests <u>solely</u> on (29.1) and (29.2). Lemma 1 amounts to saying that any $\mathcal{F}_{\sigma\delta}$ belonging to $\mathcal{C}$ satisfies (29.3), and Lemma 2 to the fact that the class $\mathcal{C}'$ in $E \times F$ consisting of the sets whose projection on F belongs to $\mathcal{C}$ still satisfies (29.1) and (29.2). Then Lemma 1 is applied in $E \times F$, and finally projection and intersection commute thanks to the compactness of the paving $\mathcal{E}$ (no. 6).

Sion calls such a class $\mathcal{C}$ satisfying (29.1) and (29.2) a <u>capacitance</u>. The validity of (29.3) then is "Sion's Capacitability theorem", which is a little bit more general than that of Choquet. See Sion [1].

Construction of capacities

The hypotheses of Choquet's theorem are quite general, but difficult to fulfill : one seldom comes across non trivial set functions which are given from start for all subsets of a set F. It is more natural to consider a function defined on a paving and to determine whether one can extend it to the whole of $\mathfrak{P}(F)$ as a Choquet capacity. Still following Choquet, we now describe such an extension procedure for "strongly subadditive" set functions. We limit ourselves to the positive case, but this restriction is by no means essential.

30 DEFINITION. Let $\mathfrak{F}$ be a paving on a set F, closed under $(\cup f, \cap f)$. Let I be a positive and increasing set function defined on $\mathfrak{F}$(1). We say that I is strongly sub-additive if for every pair (A,B) of elements of $\mathfrak{F}$

(30.1) $$I(A \cup B) + I(A \cap B) \leq I(A) + I(B).$$

If the symbol $\leq$ is replaced by =, we get the definition of an additive function on $\mathfrak{F}$.

31 THEOREM. Let $\mathfrak{F}$ be a paving on F which is closed under $(\cup f, \cap f)$ and let I be an increasing and positive set function on $\mathfrak{F}$. The following properties are equivalent :

(a) I is strongly subadditive ;

(b) $I(P \cup Q \cup R) + I(R) \leq I(P \cup R) + I(Q \cup R)$ for all $P,Q,R \in \mathfrak{F}$;

(c) $I(Y \cup Y') + I(X) + I(X') \leq I(X \cup X') + I(Y) + I(Y')$ for all pairs (X,Y), (X',Y') of elements of $\mathfrak{F}$ such that $X \subset Y$, $X' \subset Y'$.

Proof : To show that (a) $\Rightarrow$ (b), we write $A = P \cup R$, $B = Q \cup R$ in (30.1). Then

$$I(P \cup Q \cup R) + I((P \cap Q) \cup R) \leq I(P \cup R) + I(Q \cup R).$$

Since I is increasing, the inequality implies (b).

To show that (b) $\Rightarrow$ (c), we write $P = Y$, $Q = Y'$, $R = X$ in (b). Then :

$$I(Y \cup Y' \cup X) + I(X) \leq I(Y \cup X) + I(Y' \cup X).$$

We add I(X') to both sides and use the relations

$$Y \cup Y' \cup X = Y \cup Y', \quad Y \cup X = Y, \quad Y' \cup X = Y' \cup X \cup X'.$$

Then

$$I(Y \cup Y') + I(X) + I(X') \leq I(Y) + [I(Y' \cup X \cup X') + I(X')].$$

We again apply (b) with $P = Y'$, $Q = X$, $R = X'$. So we get an upper bound for the bracket to the right of the preceding inequality and deduce

$$I(Y \cup Y') + I(X) + I(X') \leq I(Y) + I(Y' \cup X') + I(X \cup X')$$

$$= I(Y) + I(Y') + I(X \cup X').$$

(1) The value $+\infty$ is allowed.

That is (c).

To show (c) $\Rightarrow$ (a), it suffices to write $X = A \cap B$, $Y = B$, $X' = Y' = A$ in (c). Then

$$I(A \cup B) + I(A \cap B) + I(A) \leq I(A) + I(B) + I(A).$$

Then either $I(A) = +\infty$ and inequality (30.1) is trivial, or $I(A) < +\infty$ and this inequality implies (30.1). Thus the equivalence of all three properties is established.

REMARKS. Formula (c) extends immediately by induction as follows : Let $X_1, X_2, \ldots, X_n, Y_1, Y_2, \ldots, Y_n$ be elements of $\mathcal{F}$ such that $X_i \subset Y_i$ for $i = 1,2,\ldots, n$. Then

(31.1) $$I(\bigcup_i Y_i) + \sum_i I(X_i) \leq I(\bigcup_i X_i) + \sum_i I(Y_i).$$

This formula looks more pleasant when all the quantities $I(X_i)$ are finite ; it can then be written :

(31.2) $$I(\bigcup_i Y_i) - I(\bigcup_i X_i) \leq \sum_i [I(Y_i) - I(X_i)].$$

Inequality (b) is less useful ; when all the quantities appearing are finite, it can be written as

$$I(P \cup Q \cup R) - I(P \cup R) - I(Q \cup R) + I(R) \leq 0.$$

We now associate an 'outer capacity" to every strongly subadditive and increasing set function, and investigate whether this procedure yields a true Choquet capacity.

THEOREM. Let F be a set with a paving $\mathcal{F}$ closed under $(\cup f, \cap f)$. Let I be a set function 32
defined on $\mathcal{F}$, positive, increasing and strongly subadditive, which satisfies the following property :

(32.1) for every increasing sequence $(A_n)_{n \geq 1}$ of elements of $\mathcal{F}$ whose union A belongs to $\mathcal{F}$, $I(A) = \sup_n I(A_n)$.

For every set $A \in \mathcal{F}_\sigma$ we define

(32.2) $$I^*(A) = \sup_{\substack{B \in \mathcal{F} \\ B \subset A}} I(B).$$

and, for every subset C of F :

(32.3) $$I^*(C) = \inf_{\substack{A \in \mathcal{F}_\sigma \\ A \supset C}} I^*(A) \quad (1) \qquad (\inf \emptyset = +\infty).$$

Then the function I^* is increasing and has the following properties :

(a) for every increasing sequence $(X_n)_{n \geq 1}$ of subsets of F,

(32.4) $$I^*(\bigcup_n X_n) = \sup_n I^*(X_n).$$

(b) let (X_n), (Y_n) be two sequences of subsets of F such that $X_n \subset Y_n$ for all n. Then :

(32.5) $$I^*(\bigcup_n Y_n) + \sum_n I^*(X_n) \leq I^*(\bigcup_n X_n) + \sum_n I^*(Y_n).$$

---

(1) I^* is called the outer capacity associated with I.

(c) <u>The function</u> I^* <u>is an</u> $\mathcal{F}$-<u>capacity</u>, <u>if and only if</u>

$$I^*(\bigcap_n A_n) = \inf_n I(A_n) \tag{32.6}$$

<u>for every decreasing sequence</u> $(A_n)_{n \geq 1}$ <u>of elements of</u> $\mathcal{F}$.

<u>Proof</u> : We start by noting that definition (32.2) gives an extension of I to $\mathcal{F}_\sigma$ and definition (32.3) an extension of I^* to the whole of $\mathcal{P}(F)$. In other words, the definition of I^* is coherent. Clearly I^* is increasing on $\mathcal{P}(F)$.

(1) <u>Let</u> $(A_n)_{n \geq 1}$ <u>be an increasing sequence of elements of</u> $\mathcal{F}_\sigma$ <u>and set</u> $A = \bigcup_n A_n$. <u>Then</u> $I^*(A) = \sup_n I^*(A_n)$.

It obviously suffices to show that $I(B) \leq \sup I^*(A_n)$ for all $B \in \mathcal{F}$ such that $B \subset A$. Let $(A_{nm})_{m \geq 1}$ be a sequence of elements of $\mathcal{F}$ whose union is A_n. Replacing if necessary each A_{nm} by the set

$$A'_{nm} = A_{1m} \cup A_{2m} \cup \ldots \cup A_{nm},$$

we can assume that A_{nm} is an increasing function of n for each m. Then

$$\sup I^*(A_n) = \sup_n(\sup_m I(A_{nm})) = \sup_n I(A_{nn}).$$

Let B be an element of $\mathcal{F}$ contained in A ; then $B = \bigcup_n (B \cap A_{nn})$ and, by (32.1)

$$I(B) = \sup_n I(B \cap A_{nn}) \leq \sup_n I(A_{nn}) = \sup_n I^*(A_n).$$

(2) <u>The function</u> I^* <u>is strongly subadditive on</u> $\mathcal{P}(F)$.

First let A and B be two elements of $\mathcal{F}_\sigma$ and let (A_n), (B_n) be two increasing sequences of elements of $\mathcal{F}$ whose unions are equal respectively to A and B. Then the sets $A_n \cap B_n$, $A_n \cup B_n$ belong to $\mathcal{F}$ and

$$A \cap B = \bigcup_n (A_n \cap B_n) \; ; \; A \cup B = \bigcup_n (A_n \cup B_n) \; ;$$

hence by (1)

$$I^*(A \cup B) + I^*(A \cap B) = \lim_n I(A_n \cup B_n) + \lim_n I(A_n \cap B_n)$$
$$\lim_n [I(A_n) + I(B_n)] = I^*(A) + I^*(B).$$

We then consider any two subsets X and Y of F. Let A and B denote elements of $\mathcal{F}_\sigma$ containing X and Y respectively. We have

$$I^*(X \cup Y) + I^*(X \cap Y) \leq I^*(A \cup B) + I^*(A \cap B) \leq I^*(A) + I^*(B).$$

Then we get the desired inequality

$$I^*(X \cup Y) + I^*(X \cap Y) \leq I^*(X) + I^*(Y)$$

by passing to the infimum over A and B.

(3) <u>Let</u> $(X_n)_{n \geq 1}$ <u>be an increasing sequence of subsets of</u> F <u>and let</u> $X = \bigcup_n X_n$. <u>Then</u> $I^*(X) = \sup_n I^*(X_n)$.

If the right-hand side is $+\infty$, (3) is obvious, so we may assume it is finite. Let h be a number > 0 ; we are going to construct an increasing sequence $(Y_n)_{n \geq 1}$ of elements of $\mathcal{F}_\sigma$ such that $Y_n \supset X_n$ and

$$I^*(Y_n) \leq I^*(X_n) + h.$$

Then if Y denotes the union of the Y_n, which belongs to $\mathcal{F}_\sigma$ and contains X, we have by (1)

$$I^*(X) \le I^*(Y) = \sup_n I^*(Y_n) \le \sup_n I^*(X_n) + h$$

and the theorem will be established, since h is arbitrary.

We begin by choosing for each n a set $Z_n \in \mathcal{F}_\sigma$ such that $X_n \subset Z_n$ and $I^*(X_n) \le I^*(Z_n) \le I^*(X_n) + \frac{h}{2^n}$. We then write

$$Y_n = Z_1 \cup Z_2 \cup \ldots \cup Z_n$$

and prove inductively that

$$I^*(X_n) \le I^*(Y_n) \le I^*(X_n) + h(1-\frac{1}{2^n}),$$

which implies the required property.

Since these inequalities are obviously satisfied for n = 1, we assume they hold up to step n. We have $Y_{n+1} = Y_n \cup Z_{n+1}$; strong subadditivity implies

$$I^*(Y_{n+1}) \le I^*(Z_{n+1}) + [I^*(Y_n) - I^*(Y_n \cap Z_{n+1})].$$

Now the bracket is no greater than $h(1-\frac{1}{2^n})$, since $Y_n \cap Z_{n+1}$ is an element of $\mathcal{F}_\sigma$ lying between X_n and Y_n and we have, by the induction hypothesis,

$$I^*(X_n) \le I^*(Y_n \cap Z_{n+1}) \le I^*(X_n) + h(1-\frac{1}{2^n}).$$

Consequently

$$I^*(X_{n+1}) \le I^*(Y_{n+1}) \le I^*(Z_{n+1}) + h(1-\frac{1}{2^n})$$
$$\le I^*(X_{n+1}) + h(\frac{1}{2^{n+1}} + 1 - \frac{1}{2^n})$$

where the last inequality follows from the definition of Z_{n+1}. Hence the induction formula is true at step n + 1 and property (3) is established.

It only remains to prove (32.5) . This inequality is deduced immediately on passing to the limit from the relation

$$I^*(\bigcup_{i=1}^{n} Y_i) + \sum_{i=1}^{n} I^*(X_i) \le I^*(\bigcup_{i=1}^{n} X_i) + \sum_{i=1}^{n} I^*(Y_i)$$

which is a consequence of property (2) (see (31.1)) (the passage to the limit is justified by property (3)).

Finally, it is immediate that condition (32.6) is necessary and sufficient for I^* to be a Choquet $\mathcal{F}$-capacity.

> Assertion (6) is somewhat different from the other ones. There is no reason why I^* should be a capacity <u>relative to the same paving</u> $\mathcal{F}$ from which the extension started. For example, we shall apply Theorem 32 with $\mathcal{F}$ being either the paving of compact subsets of a Hausdorff space E or that of open subsets ; (6) will be a natural condition in the first case, but not in the second one.

Applications to measure theory

Before pursuing the study of capacities, let us show that theorems 28 and 32 contain several important and classical results of measure theory.

33 (a) Measurability of analytic sets

Let $(\Omega,\mathcal{A},\mathbb{P})$ be a complete probability space and let $\mathcal{F}$ be a family of subsets of Ω, contained in $\mathcal{A}$ and closed under $(\cup f,\cap f)$. Let I be the restriction of $\mathbb{P}$ to $\mathcal{F}$. Obviously $I^*(A) = \mathbb{P}(A)$ for every element A of $\mathcal{F}_\sigma$ and consequently also $I^*(A) = \mathbb{P}(A)$ for every element A of $\mathcal{F}_\delta$ by (32.3). Conditions (32.1) and (32.6) are obviously satisfied.

Let A be an $\mathcal{F}$-analytic subset of Ω. Choquet's theorem implies that

$$\sup_{\substack{B\in\mathcal{F}_\delta\\ B\subset A}} \mathbb{P}(B) = \inf_{\substack{C\in\mathcal{F}_\sigma\\ C\supset A}} \mathbb{P}(C).$$

So there exist an element B' of $\mathcal{F}_{\delta\sigma}$ and an element C' of $\mathcal{F}_{\sigma\delta}$ such that $B' \subset A \subset C'$ and $\mathbb{P}(B') = \mathbb{P}(C')$. This implies in particular that $A \in \mathcal{A}$. This result was known long before Choquet's theorem (see Saks [1], p. 50).

(b) Carathéodory's extension theorem

34 We return to the hypotheses of 32 and suppose that I is additive on $\mathcal{F}$(cf. 30) and that (32.6) holds. Let $(A_n)_{n\in\mathbb{N}}$ and $(B_n)_{n\in\mathbb{N}}$ be two decreasing sequences of elements of $\mathcal{F}$. Passing to the limit (according to (32.6)) in the formula

$$I(A_n \cup B_n) + I(A_n \cap B_n) = I(A_n) + I(B_n)$$

we see that I^* is additive on $\mathcal{F}_\delta$. Then let A and B be two elements of $\mathcal{A}(\mathcal{F})$ and ε a number > 0 ; we choose two sets A' and B', belonging to $\mathcal{F}_\delta$, contained respectively in A and B and such that :

$$I^*(A') \geq I^*(A) - \varepsilon \;;\; I^*(B') \geq I^*(B) - \varepsilon.$$

Then we have :

$$\begin{aligned} I^*(A \cup B) + I^*(A \cap B) &\geq I^*(A' \cup B') + I^*(A' \cap B') = I^*(A') + I^*(B') \\ &\geq I^*(A) + I^*(B) - 2\varepsilon. \end{aligned}$$

Since the function I^* is strongly sub-additive and ε is arbitrary, we see that I^* is additive on $\mathcal{A}(\mathcal{F})$.

Having established this, we consider a Boolean algebra $\mathcal{F}$ and on $\mathcal{F}$ an additive set function I, which is positive and finite and satifies Carathéodory's condition :

(34.1) If $A_n \in \mathcal{F}$ are decreasing and $\bigcap_n A_n = \emptyset$, then $\lim_n I(A_n) = 0$. Then obviously (32.1) is satisfied. We show that (32.6) is also satisfied. This condition can be stated as follows : if (G_n) is an increasing sequence and (F_n) a decreasing sequence of elements of $\mathcal{F}$ and $\bigcup_n G_n \supset \bigcap_n F_n$, then $\sup_n I(G_n) \geq \inf_n I(F)$. Now let $H_n = F_0 \setminus F_n \in \mathcal{F}$; the H_n are increasing and $\bigcup_n (G_n \cup H_n) \supset F_0$. By (32.1), $\sup_n I(G_n \cup H_n) \geq I(F_0)$ and a fortiori $\sup_n(I(G_n) + I(H_n)) \geq I(F_0)$, whence subtracting $\sup_n I(G_n) \geq \inf_n(I(F_0) - I(H_n)) = \inf_n I(F_n)$.

Hence we can apply 32 and the remark at the beginning of 34 to see that I^* is additive on $\mathcal{A}(\mathcal{F})$ and hence also on $\sigma(\mathcal{F}) \subset \mathcal{A}(\mathcal{F})$. Since I^* passes to the limit along increasing sequences, I^* is a measure on $\sigma(\mathcal{F})$ which extends I and we have established the classical Carathéodory extension theorem from probability theory.

Let us establish similarly the other main extension theorem.

(c) <u>Daniell's theorem</u> 35

Let Ω be a set and $\mathcal{H}$ a linear space of real valued functions on Ω which is closed under the operation $\wedge$ and contains the constant functions. Let λ be a linear functional on $\mathcal{H}$ which is increasing (i.e. positive on the cone $\mathcal{H}^+$ of positive elements of $\mathcal{H}$) and satisfies Daniell's condition :

(35.1) <u>for every decreasing sequence</u> (h_n) <u>of elements of</u> $\mathcal{H}^+$ <u>such that</u> $\lim_n h_n = 0$, $\lim_n \lambda(h_n) = 0$.

Let us prove <u>Daniell's theorem</u> : there exists a positive measure μ on the σ-field $\sigma(\mathcal{H})$ (unique according to I.22, the lattice form of the monotone class theorem) such that $\lambda(h) = \int h\mu$ for all $h \in \mathcal{H}$.

Let F be the set $\mathbb{R}_+ \times \Omega$. We associate to every positive function g on Ω the set $W_g = \{(t,\omega) \in F : t < g(\omega)\}$ of all points of F lying strictly below the graph of g. The mapping $g \mapsto W_g$ is injective. We denote by $\mathcal{F}$ the paving on F consisting of all W_h, $h \in \mathcal{H}^+$, which is closed under $(\cup f, \cap f)$ according to the relations

$$W_f \cup W_g = W_{f\vee g},\quad W_f \cap W_g = W_{f\wedge g}.$$

Let I be the set function defined on $\mathcal{F}$ by

$$I(W_h) = \lambda(h) \qquad\qquad (h \in \mathcal{H}^+).$$

The relation $h_1 \wedge h_2 + h_1 \vee h_2 = h_1 + h_2$ implies that I is additive on $\mathcal{F}$. Daniell's condition implies (32.1). Hence we may use the extension Theorem 32 to define a set function I^* on $\mathcal{B}(F)$. For every positive function g on Ω, we set

$$\lambda^*(g) = I^*(W_g)$$

We show that I^* satisfies (32.6). The verification reduces to that of the following statement : <u>let</u> (f_n) <u>be a decreasing sequence of elements of</u> $\mathcal{H}^+$ <u>and</u> (g_n) <u>an increasing sequence of elements of</u> $\mathcal{H}^+$ <u>such that</u> $\sup_n g_n \geq \inf_n f_n$. <u>Then</u> $\lambda^*(\sup_n g_n) \geq \inf_n \lambda(f_n)$. Let $h_n = f_0 - f_n$. These functions increase with n and belong to $\mathcal{H}^+$. The relation $\sup_n (g_n + h_n) \geq f_0$ implies $\lambda^*(\sup_n(g_n+h_n)) \geq \lambda(f_0)$ and, by (32.4), $\sup_n \lambda(g_n+h_n) \geq \lambda(f_0)$. Now $\sup_n \lambda(g_n+h_n) = \sup_n(\lambda(g_n) + \lambda(h_n)) = \sup_n \lambda(g_n) + \sup_n \lambda(h_n)$. Hence $\sup_n \lambda(g_n) \geq \lambda(f_0) - \sup_n \lambda(h_n) = \inf_n \lambda(f_n)$, the required result.

Finally, λ^* is positively homogeneous : if g is a function ≥ 0 on Ω and a is a real ≥ 0, then $\lambda^*(ag) = a\lambda^*(g)$. This property is indeed true for $g \in \mathcal{H}^+$ and is clearly preserved by the extension operations (32.2) and (32.3).

We show that W_{aI_A} belongs to $\mathcal{A}(\mathcal{F})$ for all $A \in \sigma(\mathcal{H})$ and all $a \geq 0$. Let $\mathcal{B}$ denote the family of subsets A of Ω such that W_{aI_A} and $W_{aI_{A^c}}$ belong to $\mathcal{A}(\mathcal{F})$ for all $a \geq 0$; this is σ-field since $\mathcal{A}(\mathcal{F})$ is closed under $(\cup c, \cap c)$. Hence to prove that $\sigma(\mathcal{H})$ is included in $\mathcal{B}$ we need only show that $\mathcal{B}$ contains all sets of the form $\{h > b\}$, with $h \in \mathcal{H}$, $b \in \mathbb{R}$. Since $\{h > b\} = \{(h-b)^+ > 0\}$, it suffices to show that $\{h > 0\}$ belongs to $\mathcal{B}$ for all $h \in \mathcal{H}^+$, and this follows from

$$W_{aI_{\{h>0\}}} = \bigcup_n W_{a((nh)\wedge 1)}, \quad W_{aI_{\{h=0\}}} = \bigcap_n W_{a(1-hn)^+}.$$

Then, for all $A \in \sigma(\mathcal{H})$, set

$$\mu(A) = I^*(W_{I_A}) = \lambda^*(I_A).$$

This set function is a bounded positive measure on $(\Omega,\sigma(\mathcal{H}))$: the additivity of μ follows from that of I^* on $\mathcal{A}(\mathcal{F})$ (34) and the σ-additivity from (32.4). We finally show that $\lambda(h) = \int h\mu$ for all $h \in \mathcal{H}^+$ and hence for all $h \in \mathcal{H}$. By (32.4), it suffices to show that $\lambda^*(g) = \int g\mu$ when g is an elementary function $\sum_1^n a_i 1_{A_i}$ where the A_i are disjoint elements of $\sigma(\mathcal{H})$ and the a_i are reals > 0. But under these conditions

$$\lambda^*(g) = \sum_1^n \lambda^*(a_i I_{A_i}) = \sum_1^n a_i \lambda^*(I_{A_i}) = \int g\mu$$

since λ^* is positively homogeneous and additive on $\mathcal{A}(\mathcal{F})$.

36 (d) The representation theorem of F. Riesz

We recall how Daniell's Theorem implies the "F. Riesz representation theorem". Let E be a compact metric space, $\mathcal{H}$ the space $\mathcal{C}(E)$ and λ an increasing linear functional on $\mathcal{H}$. By Dini's lemma (of which we shall see a more elaborate form in Chapter X [1] every decreasing sequence (h_n) of continuous functions on E, which converges pointwise to 0, converges uniformly to 0 and condition (35.1) is therefore satisfied. Thus, since $\mathcal{B}(E) = \sigma(\mathcal{H})$:

THEOREM. Every increasing linear functional λ on $\mathcal{C}(E)$ has a unique representation

(36.1) $$\lambda(f) = \int f\mu \qquad (f \in \mathcal{C}(E))$$

where μ is a bounded positive measure on E.

(e) Regularity of measures

The following theorem form a transition between capacities and the results of paragraph 3.

37 THEOREM. Let E be a compact metrizable space. For every bounded positive measure μ on E and every Borel (or more generally μ-measurable) set $B \subset E$,

(37.1) $$\mu(B) = \sup_K \mu(K)$$

where K runs through compact subsets of E contained in B.

Proof : Let μ^* be the outer measure associated with μ [2], which is a capacity relative to $\mathcal{K}$. Every $\mathcal{K}$-analytic set B is μ^*-capacitable and hence satisfies (37.1). On the other hand, $\mathcal{B}(E) \subset \mathcal{A}(\mathcal{K})$ (13). If B is μ-measurable, one can find two Borel sets B' and B" such that $B' \subset B \subset B''$ and $\mu(B''\setminus B') = 0$; (37.1) for B then follows from the same relation applied to the Borel set B'.

(1) First edition, no. X.b.

(2) By definition, $\mu^*(A) = \inf_{B \in \mathcal{B}(E),\, B \supset A} \mu(B)$ for every subset A of E.

THEOREM. The same statement is true for every space E homeomorphic to a universally measurable subspace of a compact metric space, and hence for every Lusin (in particular Polish), Souslin, or cosouslin metrizable space. 38

Proof : The first sentence is obvious from 37. The case of Lusin spaces follows from their definition (16), that of Polish spaces from 17 and that of Souslin and cosouslin spaces from 16 and 33.

> We shall see in no. 69 that this important result is also valid for certain non-metrizable spaces, which play an important role in analysis. We note a consequence : although the σ-field $\mathcal{B}(E)$ is not necessarily a Blackwell σ-field (if E is cosouslin, for example), every measure on E is carried by a countable union of metrizable compact subsets and hence by a Blackwell subspace. Hence results such as 26 can be extended, up to sets of measure zero.

Right-continuous capacities

To apply Theorem 32, it is necessary to verify hypotheses (32.1) and (32.6). That is why the usual capacities are constructed, either from a left-continuous function on open sets or from a right-continuous function on compact sets. We work with a Hausdorff space F and denote as usual by $\mathcal{G}$ the paving of open subsets of F and by $\mathcal{K}$ that of compact subsets of F.

DEFINITION. Let I be an increasing positive function defined on $\mathcal{G}$. I is said to be left-continuous if 39

(39.1) for every open set U and every real number $a < I(U)$, there exists a compact set $K \subset U$ such that $I(V) > a$ for every open set V containing K.

THEOREM. Let I be a function on $\mathcal{G}$ which is positive, increasing, left-continuous and strongly subadditive. Then I satisfies (32.1) relative to the paving $\mathcal{F} = \mathcal{G}$ and I^* is a capacity relative to $\mathcal{K}$. 40

Proof : Let (U_n) be an increasing sequence of open sets with union U and let $a < I(U)$. We choose a compact set $K \subset U$ satisfying (36.1). K is contained in one of the U_n and hence $\sup_n I(U_n) > a$ and finally $\sup_n I(U_n) \geq I(U)$, and (32.1) follows.

Let (K_n) be a decreasing sequence of compact sets with intersection K. By definition of $I^*(K)$ there exists, for every number $b > I^*(K)$, an open set U containing K such that $I(U) < b$. Then one of the K_n is contained in U and hence $\inf_n I^*(K_n) < b$, whence $\inf_n I^*(K_n) \leq I^*(K)$ and we have equality.

REMARK. The conclusion of Theorem 32 is capacitability of every $\mathcal{K}$-analytic set relative to I^*. Right-continuity implies a better result : if F is metrizable and separable, every Souslin set $S \subset F$ is capacitable. Imbedding F in a compact metrizable space c, set indeed $L(G) = I(G \cap F)$ for every open set G of C ; L is right-continuous and strongly subadditive and L^* and I^* coincide on subsets of F. On the other hand, S is $\mathcal{K}$-analytic in C (18).

If F is only assumed to be Hausdorff, the Souslin sets in Bourbaki's sense (67) are capacitable.

41 DEFINITION. Let J be a positive increasing function defined on $\mathcal{K}$. J is said to be right-continuous if

(41.1) for every compact set K and every real number $a > J(K)$, there exists an open set $V \supset K$ such that $J(L) < a$ for every compact set $L \subset V$.

42 THEOREM. (a) Let J be a function on $\mathcal{K}$ which is positive, increasing, strongly subadditive and right-continuous. Then J satisfies (32.1) relative to the paving $\mathcal{F} = \mathcal{K}$.

(b) For every open set G define

(42.1) $$J^+(G) = \sup_{\substack{K \in \mathcal{K} \\ K \subset G}} J(K)$$

and for every subset A of F

(42.2) $$J^+(A) = \inf_{\substack{G \in \mathcal{G} \\ G \supset A}} J^+(G).$$

Then $J^+|_{\mathcal{K}} = J$, $J^+|_{\mathcal{G}}$ is a function of open sets satisfying the hypothesis of 40, so that J^+ is a capacity relative to $\mathcal{K}$.

Proof : To make clearer the relation with the preceding results, denote by I the function of open sets defined in (42.1) : then $J^+ = I$ on $\mathcal{G}$ and on the whole of $\mathcal{P}(F)$ J^+ is the "outer capacity" I^* relative to the paving $\mathcal{G}$. This isn't the same as the outer capacity J^* relative to the paving $\mathcal{K}$, whose definition uses $\mathcal{K}_\sigma$ sets instead of open sets). The right-continuity of J means that $I|_{\mathcal{K}} = J$; on the other hand, the same arguments applied to I show that I is left-continuous on open sets. We show that I is strongly subadditive on $\mathcal{G}$, which implies that 40 applies to $I^* = J^+$, from which the remainder of the statement follows at once.

LEMMA. Let U and V be two open sets and K a compact set contained in $U \cup V$. There then exist two compact sets $L \subset U$, $M \subset V$ such that $K = L \cup M$.

$K \setminus U$ and $K \setminus V$ are two disjoint compact sets in a Hausdorff space ; hence they can be enclosed in two disjoint open sets P and Q, and we just set $L = K \setminus P$ and $M = K \setminus Q$.

Having established this lemma, we take two numbers $a < I(U \cap V)$ and $b < I(U \cup V)$ and,choose a compact set $H \subset U \cap V$ such that $J(H) > a$ and a compact set $K \subset U \cup V$ such that $J(K) > b$. Replacing K by $H \cup K$ if necessary, it can be assumed that $H \subset K$. By the lemma, we can write $K = L \cup M$ with $L \subset U$,and $M \subset V$ and, replacing them by $H \cup L$, $H \cup M$ if necessary, we can assume that L and M contain H. Then we have
$a + b \leq J(H) + J(K) \leq J(K \cap M) + J(L \cup M) \leq J(L) + J(M) \leq I(U) + I(V)$.
Passing to the upper bound over a and b, we get $I(U \cap V) + I(U \cup V) \leq I(U) + I(V)$, the required inequality.

REMARK. Let us compare the two capacities J^* (defined by means of the $\mathcal{K}_\sigma$) and J^+

(by means of the open sets). They are equal on $\mathcal{K}$ and hence on $\mathcal{K}_\sigma$, and for arbitrary A we have $J^*(A) = \inf J^+(B)$, where $B \in \mathcal{K}_\sigma$ contains A. It follows that $J^*(A) \geq J^+(A)$. If A is $\mathcal{K}$-analytic, then $J^*(A) = J^+(A)$ by Choquet's Theorem. Note that if F is metrizable and separable and A is Souslin in F and not contained in any $\mathcal{K}_\sigma$, then A is capacitable relative to J^+ but not necessarily relative to J^* ($J^*(A) = +\infty$).

The capacity $J^+ = I^*$ is computed "from outside, using open sets". The capacitability theorem applied to J^+ thus tells us that both ways of computing the capacity, from inside and from outside, are equivalent. It was already so in 32, but approximation by $\mathcal{K}_\sigma$ sets is less convenient than by open sets.
Theorem 28 itself may be given an analogous interpretation. Let $\mathcal{M}$ be the monotone class generated by $\mathcal{F}$, and for every set B define

(27.4) $$I^+(B) = \inf I(M) \quad \text{where } M \in \mathcal{M} \text{ contains } B.$$

Then I^+ is a capacity and coincides with I on $\mathcal{M}$, and 28 means that it coincides with I on $\mathcal{A}(\mathcal{F})$. Thus all results on capacitability are theorems on approximation both from outside and from inside.

Some applications of the theory of capacities

We have already given some applications of the capacitability theorem to measure theory. The following ones are of a different kind.

We begin with the second proof of the separation theorem for analytic sets, mentioned in no. 14.

Recall the notation : $\mathcal{F}$ is a semicompact paving on F which can be assumed to be closed under ($\cup f, \cap c$) and $\mathcal{C}$ is the closure of $\mathcal{F}$ under ($\cup c, \cap c$). Let Δ be the diagonal of $F \times F$. For every subset W of $F \times F$ we set 43

$I(W) = 1$ if every element of the product paving $\mathcal{C} \times \mathcal{C}$ which contains W intersects the diagonal

$I(W) = 0$ otherwise.

Let us prove that I is a capacity relative to the product paving $\mathcal{F} \times \mathcal{F}$. It is obviously increasing. We show that if $W \in \mathcal{F} \times \mathcal{F}$ is the union of an increasing sequence (W_n), then $I(W) = \lim_n I(W_n)$. It suffices to treat the case where $I(W_n) = 0$ for all n, which means there exist elements $C_n \times D_n$ of $\mathcal{C} \times \mathcal{C}$ containing W_n such that $C_n \cap D_n = \emptyset$. Replacing C_n by $\bigcap_{m \geq n} C_m$ and D_n by $\bigcap_{m \geq n} D_m$ if necessary, we can assume that the sequences (C_n) and (D_n) are increasing. Then $(\bigcup_n C_n) \times (\bigcup_n D_n)$ belongs to $\mathcal{C} \times \mathcal{C}$, contains W and does not meet the diagonal, hence $I(W) = 0$. Finally, consider a decreasing sequence $W_n = K_n \times L_n$ of $\mathcal{F} \times \mathcal{F}$ and its intersection $W = K \times L$; let us prove that $I(W) = \inf_n I(W_n)$. It suffices to treat the case where $I(W_n) = 1$ for all n, that is, where $K_n \times L_n \neq \emptyset$. Since the paving $\mathcal{F}$ is semicompact, we have $K \cap L \neq \emptyset$ and $L(W) = 1$.

Then let A and B be two non-separable $\mathcal{F}$-analytic sets. This means [(1)] that $I(A \times B) = 1$. By the capacitability theorem, since $A \times B$ is $\mathcal{F} \times \mathcal{F}$-analytic (9),

(1) If $A \times B \neq \emptyset$; we leave to the reader the case where A or B is empty.

there exists an element $K \times L$ of $\mathcal{F} \times \mathcal{F}$ contained in $A \times B$ such that $I(K \times L) = 1$. But then $K \cap L \neq \emptyset$, hence $A \cap B \neq \emptyset$ and the theorem is established [(1)].

We pass to a result which has important probabilistic applications.

44 Let $(\Omega,\mathcal{F})$ be a measurable space and A a subset of $\mathbb{R}_+ \times \Omega$. We write, for all $\omega \in \Omega$,

(44.1) $$D_A(\omega) = \inf\{t \in \mathbb{R}_+ : (t,\omega) \in A\}$$

(with the usual convention that $\inf \emptyset = +\infty$); the function D_A is called the debut of A.

THEOREM. Suppose that A belongs to the σ-field $\mathcal{B}(\mathbb{R}_+) \times \mathcal{F}$ (or, more generally, that A is $(\mathcal{B}(\mathbb{R}_+) \times \mathcal{F})$-analytic).

(a) The debut D_A is measurable relative to the σ-field $\hat{\mathcal{F}}$, the universal completion of $\mathcal{F}$.

(b) Let $\mathbb{P}$ be a probability law on $(\Omega,\mathcal{F})$. There exists an $\mathcal{F}$-measurable random variable T with values in $[0,\infty]$ such that

(44.2) $$T(\omega) < \infty \Rightarrow (T(\omega),\omega) \in A \quad \text{("T is a cross-section of A")}$$

(44.3) $$\mathbb{P}\{T < \infty\} = \mathbb{P}\{D_A < \infty\}$$

(in other words, T is an almost-complete cross-section of A) [(2)].

Proof : Let $r > 0$. The set $\{D_A < r\}$ is the projection on Ω of $\{(t,\omega) : t < r, (t,\omega) \in A\}$. By 13, $\{D_A < r\}$ is $\mathcal{F}$-analytic. By 33, it belongs to every completed σ-field of $\mathcal{F}$, whence assertion (a). Associate with $\mathbb{P}$ the set function $\mathbb{P}^*$ as in 32 ($\mathbb{P}^*$ is the classical "outer probability" of Carathéodory) : this is an $\mathcal{F}$-capacity, equal to $\mathbb{P}$ on $\mathcal{F}$ and even on the completed σ-field of $\mathcal{F}$(II.32,(b)). Let π be the projection of $\mathbb{R}_+ \times \Omega$ onto Ω and let I be the set function $A \mapsto \mathbb{P}^*[\pi(A)]$: I is a capacity relative to the paving $\mathcal{H}$, the closure of $\mathcal{K}(\mathbb{R}_+) \times \mathcal{F}$ under $(\cup c,\cap c)$ (28, Lemma 2). By 13, every element of the product σ-field $\mathcal{B}(\mathbb{R}_+) \times \mathcal{F}$ (or, more generally, of $\mathcal{A}(\mathcal{B}(\mathbb{R}_+) \times \mathcal{F})$) is $\mathcal{H}$-analytic and the capacitability theorem 28 implies the existence, for all $\varepsilon > 0$, of an element B of $\mathcal{H}_\delta = \mathcal{H}$ contained in A and such that $I(B) > I(A) - \varepsilon$. This can also be written $\mathbb{P}\{D_B < \infty\} > \mathbb{P}\{D_A < \infty\} - \varepsilon$. Since for all $\omega \in \Omega$ the set $B(\omega) = \{t : (t,\omega) \in B\}$ is compact, the graph of D_B in $\mathbb{R}_+ \times \Omega$ is contained in A. Then let S_ε be an $\mathcal{F}$-measurable positive random variable, equal almost everywhere [(3)] to D_B ; we write

$$T_\varepsilon(\omega) = S_\varepsilon(\omega) \text{ if } (S_\varepsilon(\omega),\omega) \in A$$
$$= +\infty \quad \text{otherwise.}$$

---

(1) For a proof of a deeper theorem along the same lines, see Dellacherie, Séminaire de Probabilités de Strasbourg vol. X, p. 580-582. (Lecture Notes in M. 511, Springer-Verlag 1976).

(2) With the notation of (b) the cross-section T of A is said to be complete if $T(\omega) < \infty$ for every ω such that $D_A(\omega) < \infty$. We prove in the appendeix a theorem on existence of complete cross-sections (81).

(3) In fact, D_B itself can be shown to be $\mathcal{F}$-measurable.

Then T_ε satisfies (44.2) and a weaker condition than (44.3) ($\mathbb{P}\{T_\varepsilon < \infty\} > \mathbb{P}\{D_A > \infty\} - \varepsilon$). Let us say (in this proof only) that, given $C \in \mathcal{B}(\mathbb{R}_+) \times \mathcal{F}$, a positive $\mathcal{F}$-measurable function S such that $(S(\omega),\omega) \in C$ for all $\omega \in \{S < \infty\}$ is a section of C with remainder $\mathbb{P}\{S = \infty , D_C < \infty\}$: By the above, C has a section with remainder $<\varepsilon$ for all $\varepsilon > 0$. We construct sections of A inductively as follows. $T_0 = +\infty$ identically. If T_n has been defined, we construct a section S_n of $A_n = A \cap \{(t,\omega) : T_n(\omega) = \infty\}$ such that $\mathbb{P}\{S_n < \infty\} \geq \frac{1}{2}\mathbb{P}\{D_{A_n} < \infty\}$, and we set $T_{n+1} = T_n \wedge S_n$, a section of A which "extends" T_n. At each step, the remainder is at most half of the proceding one. So $T = \inf_n T_n$ is a section with remainder zero, which therefore satisfies (44.2) and (44.3).

At the cost of minor modifications, this theorem is still valid if $(\mathbb{R}_+,\mathcal{B}(\mathbb{R}_+))$ is replaced by a Souslin measurable space $(S,\mathcal{S})$, which, from the measure theoretic point of view, is not distinguishable from an analytic subset of $\mathbb{R}$ (20) : 45

(a) is no longer meaningful but the projection $\pi(A)$ of A onto Ω still belongs to $\hat{\mathcal{F}}$; (b) remains true provided $\mathbb{P}\{D_A < \infty\}$ is replaced by $\mathbb{P}[\pi(A)]$ and $[0,\infty]$ by $S \cup \{\infty\}$, where "∞" is a point added to S. By way of illustration, here is a theorem on lifting measures, which we shall use later.

THEOREM. *Let* $(S,\mathcal{S})$ *be a Souslin measurable space*, $(E,\mathcal{E})$ *be a Hausdorff separable measurable space and* f *be a measurable mapping from* S *onto* E. *For every probability law* μ *on* E, *there exists a probability law* λ *on* S *such that* $\mu = f(\lambda)$.

Proof : We apply 44 with $\mathbb{P} = \mu$. The graph A of f in $S \times E$ belongs to $\mathcal{S} \times \mathcal{E}$. Hence there exists a measurable mapping g defined on an element B of $\mathcal{E}$ such that $\mu(B) = 1$, with values in S, such that $g(y) = x$ if $f(x) = y \in B$. Then it suffices to take λ to be the image law $g(\mu)$.

Note that the hypotheses imply that the measurable space E is Souslin.

3. BOUNDED RADON MEASURES

If abstract measure theory - which is the basis of probability theory - is compared to the theory of Radon measures, as developed for example in Bourbaki's book on integration, it may seem that the latter is superior to the former on four counts. These are, by order of decreasing importance :

- the existence of a good theorem on inverse (projective) limits of mesures,
- the existence of some reasonable topologies (vague, strict) on the space of measures,
- the possibility of passing to the limit along uncountable increasing families of l.s.c. functions.
- the removal of certain σ-finiteness restrictions.

The "importance" is here estimated from the probabilists' point of view. We leave aside the last point and examine the other three. We also prove, without many details, existence theorems for conditional laws and disintegration of measures. We follow

Bourbaki quite closedly throughout this paragraph : see Bourbaki [5].

Radon measures and filtering families of semicontinuous functions

46 DEFINITION. Let E be a Hausdorff topological space. A measure μ on E is called a Radon measure if

(1) every point of E has an open neighbourhood V such that $\mu(V) < +\infty$;

(2) for all $A \in \mathcal{B}(E)$,

(46.1) $$\mu(A) = \sup_{\substack{K \in \mathcal{K}(E) \\ K \subset A}} \mu(K).$$

Property (1) and (2) are called respectively local boundedness and inner regularity on tightness of the measure μ. A signed measure is said to be Radon if it is the difference of two positive Radon measures. Here we limit ourselves to positive measures and to bounded measures except in 47.

The notion of a (bounded) Radon measure has a counterpart in abstract theory : the notion of inner regular measure with respect to a compact paving. This notion seems to have some applications, but note of great importance. The first edition of this book can be consulted or the notes [1] of Pfanzagl-Pierlo.

47 REMARKS. (a) Property (1) implies that $\mu(K) < \infty$ for every compact set K. Conversely, on a locally compact space, this property implies (1). Every Radon measure on a compact space is bounded.

(b) Every element of the completed σ-field $\mathcal{B}^{\mu}$ of $\mathcal{B}(E)$ relative to μ contains a Borel set which differs from it only by a μ-negligible set. Hence we also have (46.1) for all $A \in \mathcal{B}^{\mu}$.

(c) If μ is bounded, the approximation (46.1) applied to the complement of $A \in \mathcal{B}^{\mu}$ gives (1)

(47.1) $$\mu(A) = \inf_{\substack{G \in \mathcal{G}(E) \\ G \supset A}} \mu(G).$$

Hence the measure is also "outer regular". More generally, when μ is not bounded, this is valid for all A contained in an open set U of finite measure (pass to the complement in U instead of E) or even in the union of a sequence of open sets of finite measure U_n (given $\varepsilon > 0$, choose an open set $G_n \supset A \cap U_n$ such that $\mu(G_n) \leq \mu(A \cap U_n) + \varepsilon.2^{-n}$; then $\mu(\bigcup_n G_n) \leq \mu(A) + 2\varepsilon$). If E itself is the union of such a sequence, (47.1) holds for all $A \in \mathcal{B}^{\mu}$; this is the case when E has a countable base.

We henceforth limit ourselves to bounded measures.

48 THEOREM. Let μ be a bounded Radon measure. For every positive Borel function f (more generally for f measurable relative to the completed σ-field $\mathcal{B}^{\mu}$), we have

(48.1) $\int f_{\mu} = \sup_h \int h\mu$, where h is u.s.c. and bounded with compact support and $0 \leq h \leq f$,

(1) Recall that $\mathcal{G}(E)$ is the paving of open subsets of E.

(48.2) $\int f\mu = \inf_g \int g\mu$, where g is l.s.c. and $g \geq f$.

Proof : First formula. Replacing f by $f \wedge n$ if necessary, we can suppose that f is bounded. There then exists a measurable function k taking only a finite number of values such that $k \leq f$ and $\mu(k) \geq \mu(f) - \varepsilon$ (Lebesgue approximation). We write k as a finite sum $\sum a_n I_{A_n}$, choose for each n a compact set $K_n \subset A_n$ such that $\mu(K_n) \geq \mu(A_n) - \varepsilon.2^{-n}/a_n$ and take $h = \sum a_n I_{K_n}$.

Second formula. We choose an elementary function $j \geq f$ such that $\mu(j) \leq \mu(f) + \varepsilon$ (Lebesgue approximation from above). We write j as a countable sum $\sum a_n I_{A_n}$, choose for each n an open set $G_n \supset A_n$ such that $\mu(G_n) \leq \mu(A_n) + \varepsilon.2^{-n}/a_n$ and set $g = \sum a_n I_{G_n}$.

THEOREM. Let μ be a bounded Radon measure on a Hausdorff space E. 49

(a) Let $(f_i)_{i \in I}$ be family of l.s.c. positive functions, filtering to the right, with upper envelope f. Then $\mu(f) = \sup_i \mu(f_i)$.

(b) Let $(g_i)_{i \in I}$ be a family of u.s.c. positive functions, filtering to the left, with lower envelope g. If there exists an index i such that $\mu(g_i) < +\infty$, then $\mu(g) = \inf_i \mu(g_i)$.

(c) If E is completely regular,(1) then for every positive l.s.c. function f we have $\mu(f) = \sup_c \mu(c)$, where c runs through the set C_f of positive continuous functions bounded above by f.

Proof : We first prove the particular case of (a) that concerns open sets : If an open set G is the union of a family of open sets G_i which is filtering to the right, then $\mu(G) = \sup_i \mu(G_i)$. This is obvious, as μ is regular and every compact set contained in G is contained in some G_i. Taking complements, we get the form of (b) for closed sets.

With every positive function h we associate the Lebesgue approximation truncated at 2^n :

$$h^{(n)} = 2^{-n} \sum_{k=1}^{k=2^{2n}} I_{\{h > k2^{-n}\}}.$$

If h is l.s.c., $h^{(n)}$ is a finite linear combination of indicators of open sets. Using the preceding result,

$$\mu(f) = \sup_n \mu(f^{(n)}) = \sup_n \mu(\sup_i f_i^{(n)}) = \sup_n \sup_i \mu(f_i^{(n)})$$
$$= \sup_i \sup_n \mu(f_i^{(n)}) = \sup_i \mu(f_i).$$

We pass on to (b) : since the family contains an integrable function, one may reduce to the case where all the g_i are bounded above by an integrable function h. Replacing the g_i by the $g_i I_{\{g_i \leq N\}}$, where N is chosen large enough so that

(1) For the definition and properties of completely regular spaces, see Bourbaki, Top. Gen. IX. §1, nos 5 to 7.

$\int_{\{h > N\}} h\mu < \varepsilon$, one may still reduce to the case where the g_i are bounded by N, which finally reduces to (a) by considering the l.s.c. functions $N - g_i$ (here the extension to non-bounded Radon measures is less obvious).

Finally, since E is completely regular, C_f is a family filtering to the right whose upper envelope is f (Bourbaki, Gen. Top., Chap. IX, §2, Proposition 5). Then we deduce (c) from (a).

50 REMARK. In particular, by (a) the union of all the μ-negligible open sets is a μ-negligible open set, whose complement S_μ is the smallest closed set carrying μ. S_μ is called the support of μ.

> We have deduced properties (a) and (b) from the regularity of μ . We shall see later (in the "digression" of nos. 63-68) another proof, using a property of the space E, not of the measure μ. There is nothing very deep in all this !

Tightness and inverse limits

The fundamental theorem on inverse limits, used in the construction of stochastic processes, is Kolmogorov's Theorem (in fact, Kolmogorov has rediscovered a much earlier result of Daniell). We shall see how more recent results follow very easily from it.

Kolmogorov's proof using Carathéodory's extension theorem is quite classical (see the first edition of this book, no. III.31 or Neveau [1], Theorem III.3.1, p. 78). We rather give a quicker proof (also classical) for separable metrizable spaces.

51 We use the following notation : the $E_n (n \in \mathbb{N})$ are separable metrizable spaces; F is their product, which is also separable and metrizable ; F_n is the product $E_0 \times E_1 \times \dots \times E_n$, p_n the canonical projection of F_{n+1} onto F_n and q_n the projection of F onto F_n.

THEOREM. For each n, let μ_n be an inner regular probability law on F_n. If the family (μ_n) satisfies the compatibility conditions $\mu_n = p_n(\mu_{n+1})$, there exists on F one and only one probability law μ such that $\mu_n = q_n(\mu)$ for all n, and μ is tight.

Proof : Uniqueness. If d_n is a metric on E_n defining the topology and bounded by 1, the topology on F is defined by the metric $d((x_n), (y_n)) = \sum 2^{-n} d_n(x_n, y_n)$. The balls relative to d are measurable relative to the product σ-field $\prod_n \mathcal{B}(E_n)$ on F, which is therefore identical to $\mathcal{B}(F)$.

Let $\mathcal{A}$ be the Boolean algebra consisting of the subsets of the form $q_n^{-1}(A_n)$ $(n \in \mathbb{N}, A_n \in \mathcal{B}(E_n))$. The condition $q_n(\mu) = \mu_n$ determines μ on $\mathcal{A}$; since $\mathcal{A}$ generates the product σ-field, the uniqueness follows from I.20.

Existence. Suppose first that the E_n (and hence F) are compact. Let $\mathcal{C}_f(F)$ be the subspace of $\mathcal{C}(F)$ consisting of the continuous functions g of the form $g_n \circ q_n$ $(n \in \mathbb{N}, g_n \in \mathcal{C}(F_n))$: C_f contains the constants and is closed under the operation $\wedge$.

The σ-field it generates contains $q_n^{-1}(\mathcal{B}(E_n))$ for all n and hence is equal to $\mathcal{B}(F) = \prod_n \mathcal{B}(E_n)$. The compatibility condition between the μ_n enables us to set $I(g) = \int g_n\mu_n$, independently of the representation $g = g_n \circ q_n$ of $g \in \mathcal{C}_f$. I obviously is an increasing linear functional on $\mathcal{C}_f$ such that $I(1) = 1$. By Dini's Lemma, I satisfies Daniell's condition (35.1) : hence there exists a unique measure μ on F such that $I(f) = \int f\mu$ for all $f \in \mathcal{C}_f$ and $\mu_n(g_n) = I(g_n \circ q_n) = \mu(g_n \circ q_n)$ for all $g_n \in \mathcal{C}(F_n)$. Hence $\mu_n = q_n(\mu)$ for all n (I.23).

In order to pass to the general case, we imbed each E_n in a compact metrizable space $\bar{E}_n$ and introduce the corresponding notation $\bar{F}_n$, $\bar{F}$, $\bar{p}_n$, $\bar{q}_n$. Each μ_n can be identified with its image $\bar{\mu}_n$ under the injection of F_n into $\bar{F}_n$, a measure on $\bar{F}_n$ carried (1) by F_n. By the above special case, there exists on $\bar{F}$ a measure $\bar{\mu}$ such that $\bar{\mu}_n = \bar{q}_n(\bar{\mu})$ for all n and the problem reduces to showing that $\bar{\mu}$ is carried by F. Now let $\varepsilon > 0$. For each n, we choose a compact set $K_n \subset E_n$ such that $\mu_{n+1}(E_0 \times E_1 \times \ldots \times K_n^C) < \varepsilon 2^{-n}$ (for $n = 0$, $\mu_0(K_n^C) < \varepsilon$); then we have $\bar{\mu}(\prod_{k<n} \bar{F}_k \times K_n^C \times \prod_{k>n} \bar{F}_k) < \varepsilon 2^{-n}$. Consequently, if K is the compact set $\prod_n K_n$ contained in F, we have $\bar{\mu}(K^C) \leq \sum_n \varepsilon . 2^{-n} = 2\varepsilon$ and $\bar{\mu}$ is carried by F. The measure $\bar{\mu}$ is tight on $\bar{F}$ (37) ; since it is carried to within 2ε by a compact subset of F, it is immediately verified that it still is tight on F.

We deduce from this result the general theorem on the construction of stochastic processes, also due to Kolmogorov. However the usefulness of this theorem is somewhat illusory : when the index set T is uncountable, the σ-field $(\mathcal{B}(E))^T$ is far from being rich enough.

COROLLARY. *Let* E *be a separable metrizable space,* T *be any index set and* F *be the product set* E^T *with the product* σ*-field* $\mathcal{F} = (\mathcal{B}(E))^T$. *For every finite subset* U *of* T, *let* F_U *denote the (metrizable) space* E^U *and* q_U *the projection of* F *onto* F_U, *and let* μ_U *be a tight probability law* (2) *on* F_U. 52

There exists a probability law μ *on* $(F,\mathcal{F})$ *such that* $q_U(\mu) = \mu_U$ *for every finite* $U \subset T$, *if and only if the following condition is satisfied* :

(52.1) *For every pair* (U,V) *of finite subsets such that* $U \subset V$, μ_U *is the image of* μ_V *under the projection of* F_V *onto* F_U.

The measure μ *then is unique*.

Proof : If T is countable, this theorem reduces immediately to the preceding theorem. So assume T is uncountable. Let $\mathcal{F}_D$ denote, for every countable subset D of T, the

(1) By définition, this means that $\bar{\mu}_n$ is carried by a Borel subset of $\bar{F}_n$ contained in F_n ; as μ_n is inner regular, it is carried by a countable union of compact sets contained in F_n.

(1) This tightness condition can be slightly relaxed : see the similar theorem (III.31) in the first edition of this book or Neveu [1].

σ-field generated by the coordinate mappings whose indices belong to D. The preceding theorem implies the existence of a unique measure μ_D on $\mathcal{F}_D$ such that $\mu_U = q_U(\mu_D)$ for every finite set U contained in D. On the other hand, if D and D' are countable and D is contained in D', obviously $\mu_{D'}$ induces μ_D on $\mathcal{F}_D$. Hence there exists one and only one set function μ on the <u>union</u> $\bigcup_D \mathcal{F}_D$ which induces μ_D on each $\mathcal{F}_D$. Now this union is the σ-field $\mathcal{F}$ and μ is completely additive (since every sequence of elements of $\mathcal{F}$ is already contained in some σ-field $\mathcal{F}_D$).

The following theorem may look more general than Kolmogorov's theorem, but our proof - borrowed from Bourbaki - reduces it to the latter. Note that the mappings p_n <u>are not assumed to be continuous</u> : this is a significant improvement (due to Parthasarathy) to Prokhorov's classical theorem on inverse limits.

53 We use the following notation : we consider a sequence of separable metrizable spaces F_n with tight probability laws μ_n and <u>universally measurable</u> mappings $p_n : F_{n+1} \to F_n$. We denote by F the <u>inverse limit</u> of the (F_n, p_n), that is, the subspace of the product $\prod F_n$ consisting of all sequences $(x_k)_{k \in \mathbb{N}}$ such that $x_k = p_k(x_{k+1})$ for all k, and by q_n^n the mapping of F into F_n which maps $(x_k)_{k \in \mathbb{N}}$ to x_n.

We say that the μ_n constitute an <u>inverse system</u> of laws (on the inverse system of spaces (F_n, p_n)) if $\mu_n = p_n(\mu_{n+1})$ for all n. Under this hypothesis

THEOREM. <u>There exists one and only one law</u> μ <u>on</u> F <u>such that</u> $\mu_n = q_n(\mu)$ <u>for all</u> n. μ <u>is called the</u> inverse limit <u>of the laws</u> μ_n.

<u>Proof</u> : Let F'_n denote the space $F_0 \times F_1 \times \ldots \times F_n$, p'_n the projection of F'_{n+1} onto F'_n, F' the space $\prod_n F_n$ and q'_n the projection of F' onto F'_n. For every n, we have an injection i_n of F_n into F'_n

$$x \mapsto (p_1 \circ \ldots \circ p_{n-1}(x), \ldots, p_{n-1}(x), x) ;$$

F being a subset of F', we denote by i the injection of F into F'. We have a diagram

$$\begin{array}{ccccc} F' & \xrightarrow{q'_{n+1}} & F'_{n+1} & \xrightarrow{p'_n} & F'_n \\ \uparrow i & & \uparrow i_{n+1} & & \uparrow i_n \\ F & \xrightarrow{q_{n+1}} & F_{n+1} & \xrightarrow{p_n} & F_n \end{array}$$

The space F' is separable and metrizable and hence so is the subspace F. The injection i is continuous and hence Borel, as are the mappings q'_n (projections) and q_n (restrictions of projections). Finally, the mappings i_n are universally measurable. We denote by μ'_n the measure $i_n(\mu_n)$ and by A'_n the image $i_n(F_n)$ in F'_n.

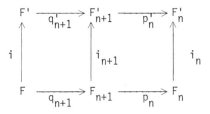

We prove that μ'_n is tight and carried by A'_n. Since i_n is universally measurable, there exists a Borel set H carrying μ_n, on which i_n coincides with a Borel function (approximate i_n by step functions) ; since μ_n is tight, there exists for all $\varepsilon > 0$ a compact set K_ε contained in H such that $\mu_n(K_\varepsilon) > 1 - \varepsilon$. The set $K'_\varepsilon = i_n(K_\varepsilon)$ is Souslin (18) and hence μ'_n-measurable (34) ; therefore $\mu'_n(K'_\varepsilon) > 1 - \varepsilon$. Further, the measure induced by μ'_n on K'_ε is tight (38). We deduce immediately that μ'_n is also tight and, the K'_ε being contained in A'_n, that μ'_n is carried by A'_n.

It follows from the diagram that $p'_n(\mu'_{n+1}) = \mu'_n$ and hence Kolgomorov's theorem implies the existence of a unique measure μ' such that $q'_n(\mu') = \mu'_n$ for all n. On the other hand, A'_n carries μ'_n and hence $q'^{-1}_n(A'_n)$ carries μ'. But this set consists of all sequences $(x_k)_{k \in \mathbb{N}}$ such that $x_0 = p_1(x_1)$ and $x_{n-1} = p_n(x_n)$. The set $\bigcap_n q'^{-1}_n(A'_n)$, which carries μ', is therefore exactly F' and the theorem follows immediately.

Narrow convergence and Prokhorov's theorem

In this section we present - following Bourbaki very closely - only the most basic results. In particular, we limit ourselves to positive measures.

DEFINITION. Let E be a completely regular space. The topology of narrow convergence[1] on the cone $\mathfrak{M}^+_b(E)$ of bounded (positive) Radon measures on E is the coarsest topology for which that mappings $\mu \mapsto \mu(f)$, where f runs through $\mathcal{C}_b(E)$, are continuous. 54

This topology is Hausdorff : for if two Radon measures μ_1, μ_2 are such that $\mu_1(f) = \mu_2(f)$ for $f \in \mathcal{C}_b$, the same property holds for all positive l.s.c. f (49,(c)) and then for all positive Borel f by (48.2). Hence $\mu_1 = \mu_2$.

Here are some elementary properties :

THEOREM. Let f be a positive l.s.c. (resp. bounded u.s.c.) function. Then the mapping $\mu \mapsto \mu(f)$ is l.s.c. (resp. u.s.c.) for the narrow topology on $\mathfrak{M}^+_b(E)$. 55

Proof : The l.s.c. case follows from 49, (c) ; if f is u.s.c. and bounded by 1, 1 - f is positive l.s.c.

Given a function f on E, the l.s.c. regularization $\underline{f}$ of f is the function $x \mapsto \liminf_{y \to x} f(y)$; this is the greatest l.s.c. function dominated by f. We define in the same way the u.s.c. regularization $\bar{f}$, and the set of points of continuity of f is the set $\{\underline{f} = \bar{f}\}$. 56

COROLLARY. Let f be a bounded Borel function on E. If the measure λ is carried by the set of points of continuity of f, the mapping $\mu \mapsto \mu(f)$ is continuous on $\mathfrak{M}^+_b(E)$ at the point λ. 57

It indeed lies between the two mappings $\mu \mapsto \mu(\underline{f})$ and $\mu \to \mu(\bar{f})$, which are equal at the point λ and respectively l.s.c. and u.s.c. on $\mathfrak{M}^+_b(E)$.

---

(1) "Narrow" convergence translates the French "convergence étroite". The usual English terminology is weak convergence, which however is slightly ambiguous when E turns out to be locally compact (in that case the true weak convergence is defined by continuous functions with compact support, the "vague convergence" of Bourbaki).

58 THEOREM. Let F be a subspace of E and let i be the injection of F into E. The mapping $\mu \mapsto i(\mu)$ is a homeomorphism of $\mathfrak{M}_b^+(F)$ onto the set of bounded positive Radon measures on E carried by F, with the topology induced by $\mathfrak{M}_b^+(E)$.

Proof : If μ is a bounded Radon measure on F, $i(\mu)$ is a measure on E carried by a countable union of compact subsets of F, according to the tightness of μ, and hence is carried by F. Conversely, if λ is a bounded Radon measure on E carried by F, λ is carried by a countable union of compact subsets of F and it follows that the measure induced by λ on F is tight. We so define two reciprocal bijections between $\mathfrak{M}_b^+(F)$ and the set $\mathfrak{M}_b^+(E,F)$ of Radon measures on E carried by F. To simplify the language, we shall identify these two sets. We must show that the two narrow topologies from E and from F coincide on $\mathfrak{M}_b^+(E,F)$. We argue with sequences, but everything extends to arbitrary filters.

First, if the $\mu_n \in \mathfrak{M}_b^+(E,F)$ converge to $\mu \in \mathfrak{M}_b^+(E,F)$ narrowly in F and if f belongs to $\mathcal{C}_b(E)$, then $f|_F$ belongs to $\mathcal{C}_b(F)$ and therefore $\mu_n(f) = \mu_n(f|_F) \to \mu(f|_F) = \mu(f)$. Hence there is convergence in $\mathfrak{M}_b^+(E)$.

Conversely, suppose that the μ_n converge to μ in $\mathfrak{M}_b^+(E)$ and let g be a continuous function on F lying between 0 and 1. Let j and k be the functions obtained by extending g to E by the values 0 and 1 outside F and let $\bar{j}$ and $\underline{k}$ be their (respectively u.s.c. and l.s.c.) regularizations. Then on F $\underline{k} = \bar{j} = g$ and hence, by 55,

$$\mu(\underline{k}) \leq \liminf_n \mu_n(\underline{k}), \quad \mu(\bar{j}) \geq \limsup_n \mu_n(\bar{j}).$$

Since the μ_n and μ are carried by F, this can be written

$$\mu(g) \leq \liminf_n \mu_n(g), \quad \mu(g) \geq \limsup_n \mu_n(g),$$

the required result.

Before giving the consequences of theorem 58 with regard to the topology of $\mathfrak{M}_b^+(E)$, we state Prokhorov's compactness theorem. For this we must recall that the set of (positive) Radon measures of mass ≤ 1 on a compact space C is compact under strict convergence (Bourbaki, [4], Integration, Chapter III, § 1, no. 9, Proposition 15). If C is metrizable, this follows very simply from 36.

59 THEOREM. Let E be a completely regular space and let H be a subset of $\mathfrak{M}_b^+(E)$ consisting of the measures of mass ≤ 1 satisfying

(59.1) For every number $\varepsilon > 0$, there exists a compact set $K \subset E$ such that $\mu(K^c) < \varepsilon$ for all $\mu \in H$. (1)

Then the closure of H in $\mathfrak{M}_b^+(E)$ is narrowly compact.

Proof : Let $\mathcal{U}$ be an ultrafilter on H ; we show that $\mathcal{U}$ converges in $\mathfrak{M}_b^+(E)$. We set $\varepsilon_n = 1/n$ and choose some compact K_n such that $\mu(K_n^c) < \varepsilon_n$. We can suppose that the

(1) This property is called equal tightness of H.

sequence (K_n) is increasing. Let μ_n be the measure $\mu.I_{K_n}$, identified to a measure on K_n ; by the compactness result recalled above, μ_n converges strictly along $\mathcal{U}$ for all n to a measure λ_n, which we consider as a measure on E carried by K_n. Let $m \leq n$ and let f be a positive continuous function on E ; fI_{K_m} and fI_{K_n} are continuous on K_m and K_n respectively and hence

$$\lambda_m(f) = \lambda_m(fI_{K_m}) = \lim_{\mathcal{U}}\mu(fI_{K_m}) \leq \lim_{\mathcal{U}}\mu(fI_{K_n}) = \lambda_n(fI_{K_n}) = \lambda_n(f).$$

Hence $\lambda_m \leq \lambda_n$. We write $\lambda = \sup_n \lambda_n$, a measure of mass ≤ 1 on E, which is obviously tight (interchange of sup). We show that μ converges to λ. Let ε be a number > 0 and f be a continuous function on E lying between 0 and 1. We choose n so large that $\varepsilon_n < \varepsilon$ and $\lambda_n(1) \geq \lambda(1) - \varepsilon$. We then have

$$|\mu(f) - \lambda(f)| \leq |\mu_n(f) - \lambda_n(f)| + \langle \mu - \mu_n, f \rangle + \langle \lambda - \lambda_n, f \rangle$$
$$\leq |\mu_n(f) - \lambda_n(f)| + \mu(K_n^c) + \langle \lambda - \lambda_n, 1 \rangle$$
$$\leq |\mu_n(f) - \lambda_n(f)| + 2\varepsilon.$$

We conclude by noting that $\mu_n(f) \to \mu_n(f)$ along $\mathcal{U}$.

The conditions of the statement also are necessary for strict compactness when E is locally compact or Polish (Prokhorov [1]). On the other hand, every narrowly convergent sequence of tight measures on a metric space satisfies (59.1). (Le Cam [1]). And yet (Preiss [1]) property (59.1) is not a necessary condition for compactness, even on a metric space as "simple" as the space of rational numbers.

The topological space of probability laws

Let E a separable metrizable space. We are going to study in the following few numbers the space $\mathcal{P}(E)$ of tight [(1)] probability laws on E, with the topology of narrow convergence. A similar study is possible for the space of tight measures of mass ≤ 1 or for the whole cone $\mathcal{M}_b^+(E)$. 60

THEOREM. If E is separable and metrizable, so is $\mathcal{P}(E)$. If further E is compact, respectively Polish, Lusin, Souslin, cosouslin, $\mathcal{P}(E)$ has the same property.

Proof : Our starting point is the well known property that $\mathcal{P}(E)$ is metrizable and compact if E is metrizable and compact.

If E is separable and metrizable, we imbed it in a compact metrizable space C. Then $\mathcal{P}(E)$ can be identified with the subspace of $\mathcal{P}(C)$ consisting of all laws carried by E (58) ; hence it is separable and metrizable.

If f is continuous and bounded on C, the function $\mu \mapsto \mu(f)$ is continuous on $\mathcal{P}(C)$ by the definition of narrow convergence ; it is l.s.c. if f is l.s.c. and bounded (55) and a simple argument on monotone classes shows that it is Borel if f is Borel

(1) On the usual spaces every law is tight (38).

and bounded. If E is Polish, E is the intersection of a sequence (G_n) of open subsets of C (17). The set of laws with no mass on G_n is closed in $\mathcal{P}(C)$, therefore $\mathcal{P}(E)$ is a $\mathcal{G}_\delta$ in the compact metric space $\mathcal{P}(C)$ and hence is Polish[1] (Bourbaki, [5], Gen. Top., Chap. IX, § 6, no. 2). If E is Lusin, E is Borel in C (19) and hence $\mathcal{P}(E) = \{\mu \in \mathcal{P}(C) : \mu(E^c) = 0\}$ is Borel in $\mathcal{P}(C)$, compact and metrizable and finally Lusin.

Suppose that E is Souslin. Then E is analytic in C and there exist a compact metric space D and an element A of $(\mathcal{K}(D) \times \mathcal{K}(C))_{\sigma\delta}$ whose image under the projection π of $D \times C$ onto C is E. By 45, $\mathcal{P}(E)$ is the image of $\mathcal{P}(A)$ under the continuous mapping $\lambda \mapsto \pi(\lambda)$ of $\mathcal{P}(D \times C)$ into $\mathcal{P}(C)$; since $\mathcal{P}(A)$ is Borel, $\mathcal{P}(E)$ is Souslin.

The study of the case where E is cosouslin needs a little more work, and an interesting definition.

61 DEFINITION. Let $(\Omega,\mathcal{F})$ be a paved set. A positive function f on Ω is called $\mathcal{F}$-analytic if, for all $a \in \mathbb{R}^+$, the set $\{f > a\}$ belongs to $\mathcal{A}(\mathcal{F})$.

We often omit mentioning $\mathcal{F}$. It is equivalent to say that $\{f \geq a\}$ belongs to $\mathcal{A}(\mathcal{F})$ for every $a > 0$, but the sets $\{f < a\}$, $\{f \leq a\}$ are complements of analytic sets. The indicator of an analytic set is an analytic function (a remark useful in remembering which way goes the inequality!) and it is easily seen using Lebesgue approximation that f is analytic if and only if f is the limit of an increasing sequence of finite linear combinations, with positive coefficients, of indicators of elements of $\mathcal{A}(\mathcal{F})$ [2].

We now prove a lemma.

62 THEOREM. Let f be a function ≥ 0 on a compact metrizable space C. If f is analytic (resp. Borel, universally measurable), then the function $\mu \to \mu(f)$ is analytic (resp. Borel, universally measurable) on $\mathcal{P}(C)$.

Proof : The Borel case has been treated above. It suffices to treat the case where f is the indicator of a set E.

Suppose that E is analytic and consider the set A from the proof of 60. The set $\{\mu \in \mathcal{P}(C) : \mu(E) > a\}$ is empty for $a \geq 1$ and, for $a < 0$, is the image of $]a,1] \times \mathcal{P}(A) \times \mathcal{P}(C)$ under the mapping $(t,\lambda,\lambda') \mapsto t.\pi(\lambda) + (1-t)\lambda'$; hence it is analytic by 18.

We suppose finally that E is universally measurable, and let Λ be a bounded measure on $\mathcal{P}(C)$. We define λ on C by writing $\lambda(B) = \int \mu(B)\Lambda(d\mu)$ for all $B \in \mathcal{B}(C)$. Then let B_1 and B_2 be two Borel subsets of C such that $B_1 \subset E \subset B_2$ and $\lambda(B_1) = \lambda(B_2)$: then $\mu(B_1) \leq \mu(E) \leq \mu(B_2)$ for all $\mu \in \mathcal{P}(C)$ and $\int \mu(B_1)\Lambda(d\mu) = \int \mu(B_2)\Lambda(d\mu)$. It follows that $\mu \mapsto \mu(E)$ is universally measurable.

We now complete the proof of 60 : suppose that E is cosouslin. The set $\mathcal{P}(E)^c$ is the set of $\mu \in \mathcal{P}(C)$ such that $\mu(E^c) > 0$; since E^c is analytic, the function $\mu \mapsto \mu(E^c)$

(1) This can be proved by constructing explicitly a distance on $\mathcal{P}(E)$ from a distance on E (cf. Prokhorov [1], Strassen [1]).

(2) It can also be shown that f is $\mathcal{F}$-analytic if and only if the set $W_f = \{(t,\omega) \in \mathbb{R}_+ \times \Omega : t < f(\omega)\}$ is $(\mathcal{B}(\mathbb{R}_+) \times \mathcal{F})$-analytic.

is analytic and $\mathcal{P}(E)$ is the complement on an analytic set.

Digression : countability properties, non-metrizable Lusin spaces

Cartier [1] has remarked that in Bourbaki's General Topology, Chapter IX, 2 nd edition the word "metrizable" can be replaced by "Hausdorff" in every section dealing whith Souslin or Lusin spaces, and that this modification (which now appears in the "definitive" edition) is quite interesting : many spaces important in analysis, and in particular the space $\mathcal{C}_c^\infty(\mathbb{R}^n)$, and its dual $\mathcal{D}(\mathbb{R}^n)$, the space of distributions on $\mathbb{R}^n$, are non-metrizable Lusin spaces.

Given the importance that the theory of random distributions may take on in the future, we show in this section how easily Bourbaki's theory of Lusin and Souslin spaces reduces to that we have just described. Besides that, the lemmas on Lindelöf spaces which we use to this end (and which are borrowed from Bourbaki) are interesting in themselves.

DEFINITION. A Hausdorff space E is (L) if every open covering of E contains a coun- 63
table subcovering, (LL) if every open set of E is (L), (LLL) if $E \times E$ is (LL).

> (L) means "Lindelöf"; this convenient but ridiculous notation will be used only in this section.

Every space with a countable base (in particular every separable metrizable space) is (LLL). So is every Hausdorff space E with the following property : there exists an (LLL) space F and a continuous mapping f of F onto E. This will be the case for all non-metrizable spaces which we shall meet later.

THEOREM. Every family $(f_i)_{i \in I}$ of l.s.c. (u.s.c.) functions on an (LL) space E 64
contains a countable subfamily with the same upper (lower) envelope.

Proof : We treat the case of l.s.c. functions. Let $f = \sup_{i \in I} f_i$. For every real a, the union of the family of open set $\{f_i > a\}$ $(i \in I)$ is the open set $\{f > a\}$; let J_a be a countable subset of I such that $\bigcup_{i \in J_a} \{f_i > a\} = \bigcup_{i \in I} \{f_i > a\}$. We set $g = \sup_{i \in J} f_i$, where J is the union of the J_a for a rational ; then $\{f > a\} = \{g > a\}$ for every rational a, hence f = g and the theorem is established.

> An equivalent statement : for arbitrary f_i, $i \in I$, there exists a countable set J such that $\sup_{i \in I} f_i$ and $\sup_{i \in J} f_i$ have the same l.s.c. regularization. This is also true for inf (without replacing l.s.c. by u.s.c.!) if E has a countable base. This is "Choquet's Lemma", cf. Brelot 1 , p. 6.

COROLLARY. Let μ be a positive measure on E and $(f_i)_{i \in I}$ be a family of positive l.s.c. 65
functions on E which is filtering to the right and let $f = \sup_i f_i$. Then $\mu(f) = \sup_i \mu(f_i)$.

The proof is obvious. There is an analogous statement for u.s.c. functions.

THEOREM. Let E be an (LLL) space. There exists a Hausdorff topology with a countable 66
base coarser than the topology on E. If E is completely regular, this topology can

be assumed to be metrizable.

Proof : To every pair (x,y) of points of E we associate a pair (U_x, U'_y) of disjoint open sets containing respectively x and y. The complement of the diagonal Δ of $E \times E$ then is the union of the open sets $U_x \times U'_y$. By the (LLL) property, there exists a sequence of pairs (x_n, y_n) such that Δ^c is the union of the $U_{x_n} \times U'_{y_n}$. Let T be the topology generated by the sets U_x, U'_y ; T is Hausdorff, coarser than the initial topology and one checks immediately that it has a countable base.

If E is completely regular, there exists a family $(f_i)_{i \in I}$ of continuous functions with values in the interval $[0,1]$, which separates the points of E. The intersection of the closed sets $F_i = \{(x,y) : f_i(x) = f_i(y)\}$ then is the diagonal Δ. By the (LLL) property, there exists a sequence (i_n) of elements of I such that Δ is the intersection of the F_{i_n}. Then the function $d(x,y) = \sum_n 2^{-n}|f_{i_n}(x) - f_{i_n}(y)|$ is a distance on E defining a topology T' coarser than the original one. Since E has the (L) property, so does T'. Hence for all $\varepsilon > 0$, there hence exists a countable family of open balls of radius ε covering E, and we deduce that E is separable under T'.

COROLLARY. Every (LLL) compact space is metrizable.

Here now is the class of - not necessarily metrizable - topological spaces introduced by Bourbaki.

67 DEFINITION. Let E be a Hausdorff topological space. E is called Souslin (resp. Lusin if there exist a Souslin (resp. Lusin) metrizable space P and a continuous (resp. injective continuous) mapping of P onto E.

P can always be assumed to be Polish (see the Appendix) : we then recover exactly Bourbaki's definition.

Every Lusin space is Souslin ; every Souslin space is separable and (LLL). Every compact subspace of a Souslin space is (LLL) and hence metrizable.

The fundamental result on Souslin and Lusin spaces in Bourbabki's sense is the fact that, from the measure theoretic point of view, they are ordinary Souslin and Lusin spaces. But we shall also improve somewhat the theorems on direct images and isomorphisms : comparison with 18 and 21 shows that the hypothesis on f has been strengthened (continuity instead of measurability) and that on F modified (separability of the σ-field isn't assumed).

68 THEOREM. Let P and F be two Hausdorff topological spaces, f be a continuous mapping of P into F and E be the image f(P).

(a) If P is Souslin, then E is $\mathcal{B}(F)$-analytic and the measurable space $(E, \mathcal{B}(E))$ is Souslin.

(b) If P is Lusin and f is injective, then $E \in \mathcal{B}(F)$ and the measurable space $(E, \mathcal{B}(E))$ is Lusin.

In particular, if we apply this to Definition 67 with F = E, we see that the measurable space underlying a Lusin (resp. Souslin) space is Lusin (Souslin) as stated earlier. Hence all of the "measurable" theory described above applies to these spaces.

Proof : Bearing in mind Definition 67, there is no loss in generality in supposing further that P is metrizable and hence has a countable base. We first establish a lemma.

LEMMA. There exists a separable sub-σ-field $\mathcal{C}$ of $\mathcal{B}(F)$ such that every point of E is an atom of $\mathcal{C}$.

Let (U_n) be a countable base for the topology on P. The σ-field generated by the sets $\overline{f(U_n)}$ is the required sub-σ-field. For let $x \in E$; for every $y \in F$, $y \neq x$, there exists an open neighbourhood H of x such that $y \notin \bar{H}$. The open set $f^{-1}(H)$ in P contains at least one U_n and then $\overline{f(U_n)}$ contains x and not y.

Let $\mathcal{C}$ be such a σ-field and let i be the canonical mapping of F onto the Hausdorff space $(\dot{F},\dot{\mathcal{C}})$ associated with $(F,\mathcal{C})$. Then $\mathcal{C} = i^{-1}(\dot{\mathcal{C}})$ and $E = i^{-1}(i(E))$. Theorems 18 and 21 applied to $i \circ f$ show that $i(E) \in \mathcal{A}(\dot{\mathcal{C}})$ and $i(E) \in \dot{\mathcal{C}}$ if f is injective and P Lusin. Hence $E \in \mathcal{A}(\mathcal{C}) \subset \mathcal{A}(\mathcal{B}(F))$ and $i(E) \in \dot{\mathcal{C}} \subset \mathcal{B}(F)$ if f is injective and P Lusin. More generally, if f is injective and P Lusin, the direct image $i \circ f(A)$ of an element A of $\mathcal{B}(P)$ belongs to $\dot{\mathcal{C}}$; hence f(A) belongs to $\mathcal{C} \subset \mathcal{B}(F)$ and f is an isomorphism of $(P,\mathcal{B}(P))$ onto $(E,\mathcal{B}(F)|_E) = (E,\mathcal{B}(E))$. Hence the latter measurable space is Lusin.

It only remains to show that, if P is Souslin or f is not injective, $(E,\mathcal{B}(E))$ is Souslin. But we know that $(E,\mathcal{C}|_E)$ is Souslin(18) and it suffices to show that every $A \in \mathcal{B}(E)$ belongs to $\mathcal{C}|_E$. To this end we choose a Borel subset A' of F such that $A = A' \cap E$ and denote by $\mathcal{C}'$ the σ-field generated by $\mathcal{C}$ and A' ; the preceding argument applies to $\mathcal{C}'$ and consequently the space $(E,\mathcal{C}'|_E)$ is Souslin. Since the two σ-fields $\mathcal{C}|_E$ and $\mathcal{C}'|_E$ are Souslin, comparable, and have the same atoms, Blackwell's theorem implies they are equal and the proof is complete.

REMARK. An argument similar to the above yields the separation theorem in Bourbaki's form : in a Hausdorff topological space two disjoint Souslin subspaces are separable by disjoint elements of $\mathcal{B}(F)$.

THEOREM. Let E be a Souslin space in Bourbaki's sense (and a fortiori Lusin...). 69
Every bounded measure μ on $\mathcal{B}(E)$ is tight.

Proof : Let μ be a bounded measure on E and let P and f have the same meaning as in Definition 67. Since the measurable space $(E,\mathcal{B}(E))$ is Souslin and hence isomorphic to a separable metrizable space, there exists by 45 a measure λ on P such that $\mu = f(\lambda)$. Since the measure μ is the image of a tight measure (38) under a continuous mapping, it is itself tight. This result is all the more interesting since the compact subsets of E are metrizable (68).

Disintegration of measures

70 The theorem on disintegration of measures has a bad reputation, and probabilists often try to avoid the use of conditional distributions... But it really is simple and easy to prove. We shall give precise statements for future reference, and rapid proofs.

This is how the problem arises : we have two probability spaces and a measurable mapping of the first into the second one

$$(70.1) \qquad (\Omega,\mathcal{F},\mathbb{P}) \underset{q}{\rightarrow} (E,\mathcal{E},\mu).$$

We suppose that $q(\mathbb{P}) = \mu$. To disintegrate $\mathbb{P}$ consists in finding an $\mathcal{E}$-measurable family (II.13) $x \mapsto \mathbb{P}_x$ of probability laws on $(\Omega,\mathcal{F})$ such that $\mathbb{P} = \int \mathbb{P}_x \mu(dx)$ and $\mathbb{P}_x$ is carried by $q^{-1}\{x\}$ (1) for μ-almost all x.

The relation with the problem of conditional laws is the following : let f be a positive $\mathcal{E}$-measurable function and g a positive $\mathcal{F}$-measurable function. We have

$$\mathbb{E}[g.f \circ q] = \int \mathbb{E}_{\mathbb{P}_x}[g.f \circ q]\, \mu(dx) = \int f(x)\mathbb{E}_{\mathbb{P}_x}[g]\mu(dx)$$

since (for μ-almost all x) $f \circ q$ is equal $\mathbb{P}_x$ - a.e. to the constant $f(x)$. This means that $\mathbb{E}_{\mathbb{P}_x}[g]$ can be interpreted as the conditional expectation of g "given that q = x" (II.38) and that $\omega \mapsto \int g\mathbb{P}_{q(\omega)}$ is a version of the conditional expectation $\mathbb{E}[g|\sigma(q)]$.

We shall show that if $(\Omega,\mathcal{F},\mathbb{P})$ is a "good" probability space, then the problem of conditional laws can be solved in a satisfactory way and that a small hypothesis on E then enables us to complete the disintegration.

71 <u>First case</u> : Ω <u>is a compact metric space with its Borel</u> σ-<u>field</u> (2) $\mathcal{B}(\Omega) = \mathcal{F}$.

To every function $f \in \mathcal{C}(\Omega)$ we associate the (bounded, not necessarily positive) measure $q(f.\mathbb{P})$ on $(E,\mathcal{E})$. This measure is absolutely continuous with respect to μ and hence admits (Radon-Nikodym Theorem) an $\mathcal{E}$-measurable density d_f, which we choose arbitrarily within its class.

Let $\mathcal{H} \subset \mathcal{C}(\Omega)$ be a countable vector space over the field $\mathbb{Q}$ of rationals, which is closed under the operations $\wedge$ and $\vee$, contains the function 1 and is dense in $\mathcal{C}(K)$. Let A be the set of all $x \in E$ such that $f \mapsto d_f(x)$ is an increasing $\mathbb{Q}$-linear functional <u>on the space</u> $\mathcal{H}$ such that $d_1(x) = 1$. It is immediately verified that A belongs to $\mathcal{E}$ and that $\mu(A) = 1$. If $x \in A$, the linear functional $d_f(x)$ can be extended to an increasing linear functional of norm 1 on $\mathcal{C}(\Omega)$, that is, a probability law on Ω(36). We denote this law by $\mathbb{P}_x$ and the corresponding expectation by $\mathbb{E}_x$. On the other hand we choose any law θ on Ω and set $\mathbb{P}_x = \theta$ if $x \notin A$.

The function $x \mapsto \mathbb{E}_x[f]$ is $\mathcal{E}$-measurable if $f \in \mathcal{H}$, hence also if $f \in \mathcal{C}(\Omega)$ by

---

(1) This condition is natural only if the atoms of $\mathcal{E}$ are the points of E.

(2) These results can be extended to the σ-field $\mathcal{F} = \mathcal{B}_u(\Omega)$.

uniform convergence, and finally if f is $\mathcal{F}$-measurable and bounded, by a simple argument using monotone classes.

We verify that, if f is $\mathcal{F}$-measurable and bounded or positive on Ω, the mapping $\omega \mapsto \mathbb{E}_{q(\omega)}[f]$ is a version of the conditional expectation $\mathbb{E}[f|\sigma(q)]$. It suffices to verify this when $f \in \mathcal{H}$. This function is of the form $h \circ q$, where $h = \mathbb{E}[f]$ is $\mathcal{E}$-measurable ; hence it is $\sigma(q)$-measurable. Conversely (I.18), every bounded $\sigma(q)$-measurable r.v. can be written as $g \circ q$, where g is $\mathcal{E}$-measurable and bounded. The fundamental property of conditional expectations therefore reduces to the equality (to be verified)

(71.1) $$\int f(\omega)g(q(\omega))\mathbb{P}(d\omega) = \int \mathbb{E}_{q(\omega)}[f]g(q(\omega))\mathbb{P}(d\omega).$$

The left-hand side is the integral of g with respect to the measure $q(f.\mathbb{P})$ and hence its value is $\int g(x)d_f(x)\mu(dx)$ by definition of d_f. The right-hand side can be written as $\int \mathbb{E}_x[f]g(x)\mu(dx)$ by definition of the image law μ and equality follows from the fact that $d_f(x) = \mathbb{E}_x[f]_\mu$- a.e.

Second case : Ω is a separable metrizable space and $\mathcal{F} = \mathcal{B}(\Omega)$ [(1)] ; the measure $\mathbb{P}$ is tight. 72

Then $\mathbb{P}$ is carried by a set J which is a countable union of compact subsets of Ω . We imbed Ω in a compact metric space K, we identify $\mathbb{P}$ with a measure on K carried by the (Borel) subset J and we construct the $\mathbb{P}_x$ on K as above. We have $\mathbb{P}(J^c) = 0$ and hence the $\mathcal{E}$-measurable set $\{x : \mathbb{P}_x(J^c) = 0\}$ carries μ. We modify $\mathbb{P}_x$ outside this set, giving it the value θ , an arbitrary law on J. All the laws $\mathbb{P}_x$ are then carried by J and hence by Ω and we can forget about the compactification.

> If Ω is homeomorphic to a universally measurable subspace of a compact metric space, this applies to every law $\mathbb{P}$ (38). Similarly, if Ω is a Bourbaki Lusin or Souslin space, we can "lift" $\mathbb{P}$ to a metrizable Souslin space P above Ω(67) by means of a section, disintegrate in P and then go down again to Ω. So this theorem covers the usual needs of analysis.

Until now the space $(E,\mathcal{E})$ has been an abstract one : hence we can take $\mathcal{E}$ to be a sub-σ-field of $\mathcal{F}$ with q the identity mapping. This special case deserves being stated :

THEOREM. Let Ω be a separable metrizable space with its Borel σ-field $\mathcal{F} = \mathcal{B}(\Omega)$ and let $\mathbb{P}$ be a tight law on $\mathcal{F}$. There then exist conditional laws on Ω relative to any sub-σ-field $\mathcal{E}$ of $\mathcal{F}$.

Moreover the same result holds for a sub-σ-field $\mathcal{E}$ of the completed σ-field $\mathcal{F}^{\mathbb{P}}$, but we omit the details. Clearly the conditional laws here are not carried by the $q^{-1}\{y\}$!

We now come back to the problem raised in 70 and wish to examine whether the laws $\mathbb{P}_x$ are carried by the $q^{-1}\{x\}$ for μ-almost all x. We have remarked that this needs 73

(1) These results can be extended to the σ-field $\mathcal{F} = \mathcal{B}_u(\Omega)$.

hypothesis on $(E,\mathcal{E})$. Hence we require, in addition to 72, the following property :

(73.1) $\quad$ E is a separable metrizable space and $\mathcal{E} = \mathcal{B}(E)$.

Let G be the product space $\Omega \times E$ with its Borel σ-field $\mathcal{G} = \mathcal{B}(G) = \mathcal{F} \times \mathcal{E}$. Let p be the mapping $\omega \mapsto (\omega, q(\omega))$ of Ω into G. One sees easily that

$$\text{the image law of } \mathbb{P} \text{ under } p \text{ is the integral } \int_E \mathbb{P}_x \otimes \varepsilon_x \mu(dx)$$

(compare these two laws on rectangular sets). Let J be a countable union of compact subsets of Ω carrying $\mathbb{P}$; J is obviously Lusin, so $p(J)$ is Souslin (18) and hence universally measurable in G, and finally it carries the image law. We deduce that for μ-almost all x, $\mathbb{P}_x$ is carried by the section J_x of $p(J)$ by x and this is contained in $q^{-1}\{x\}$.

Bimeasures

We give a last result concerning tightness, which is nice, and important in some applications. The proof is borrowed from Morando [1].

74 THEOREM. Let E and F be two separable metrizable spaces and β a mapping of $\mathcal{B}(E) \times \mathcal{B}(F)$ into the interval $[0,1]$ such that $\beta(E \times F) = 1$. Assume β has the following property

for all $A \in \mathcal{B}(E)$, $\beta(A,.)$ is a tight measure on F and, for all $B \in \mathcal{B}(F)$, $\beta(.,B)$ is a tight measure on E.

Then there exists a unique probability law $\mathbb{P}$ on $E \times F$ such that $\beta(A,B) = \mathbb{P}(A \times B)$ $(A \times \mathcal{B}(E),\ B \in \mathcal{B}(F))$ and $\mathbb{P}$ is tight.

Proof : Let $\mathcal{A}$ be the Boolean algebra consisting of finite unions of "rectangular" sets, i.e. sets of the form $A \times B$, $A \in \mathcal{B}(E)$, $B \in \mathcal{B}(F)$. Every element U of $\mathcal{A}$ also is a finite union of disjoint rectangular sets $A_i \times B_i$ $(1 \le i \le n)$ and one can readily check that the number $\sum_{i=1}^{n} \beta(A_i, B_i)$ depends only on U and not on its decomposition. We denote it by $\mathbb{P}(U)$. The function $U \mapsto \mathbb{P}(U)$ on $\mathcal{A}$ is obviously additive.

Keeping the above notation, let $\varepsilon > 0$ be given. Since the measure $\beta(A_i,.)$ is tight, there exists a compact set $T_i \subset B_i$ such that $\beta(A_i,T_i) \ge \beta(A_i,B_i) - \varepsilon/2n$. Since the measure $\beta(.,T_i)$ is tight, there exists a compact set $S_i \subset A_i$ such that $\beta(S_i,T_i) \ge \beta(A_i,T_i) - \varepsilon/2n$. If K denotes the compact set $\bigcup_i S_i \times T_i$, which belongs to $\mathcal{A}$, then $K \subset U$ and $\mathbb{P}(K) \ge \mathbb{P}(U) - \varepsilon$.

Next consider a decreasing sequence (U_n) of elements of $\mathcal{A}$, whose intersection is empty. To show that $\mathbb{P}(U_n) \to 0$, let us assume that $\mathbb{P}(U_n) \ge 3a > 0$ and deduce a contradiction. For all n, let K_n be a compact set belonging to $\mathcal{A}$, contained in U_n and such that $\mathbb{P}(U_n \setminus K_n) \le a2^{-n}$; if we set $L_n = K_0 \cap \ldots \cap K_n$, then $\mathbb{P}(U_n \setminus L_n) \le 2a$, hence $\mathbb{P}(L_n) \ge a$ and L_n is non-empty. Hence the intersection of the decreasing sequence of compact sets (L_n) is non-empty and finally $\bigcap_n U_n \ne \emptyset$, which contradicts the hypothesis.

Carathéodory's extension theorem (34) then enables us to extend $\mathbb{P}$ to a probability law on $\sigma(\mathcal{A}) = \mathcal{B}(E \times F)$. The tightness of this law is verified using monotone classes, since the property is true on $\mathcal{A}$: there is no difficulty with increasing sequences and the argument for decreasing sequences is similar to that we have just given.

CHAPTER IV

Stochastic processes

In the first two paragraphs of this chapter we study stochastic processes and methods leading to the construction of suitable versions of them. In the last two paragraphs the fundamental structure is that of a probability space provided with an increasing family of σ-fields. The study is pushed as far as we can without martingale theory.

1. GENERAL DEFINITIONS ON PROCESSES

Definition of processes

DEFINITION. *Let* $(\Omega, \mathcal{F}, \mathbb{P})$ *be a probability space*, $\mathbb{T}$ *be any set and* $(E, \mathcal{E})$ *be a measurable space. A* stochastic process *(or simply a process) defined on* Ω, *with* time set $\mathbb{T}$ *and* state space E, *is any family* $(X_t)_{t \in \mathbb{T}}$ *of* E-*valued random variables, indexed by* $\mathbb{T}$. 1

The space Ω is often called the *sample space* of the process, and the random variable X_t is called the *state at time* t. For every $\omega \in \Omega$, the mapping $t \mapsto X_t(\omega)$ from $\mathbb{T}$ into E is called the *(sample) path* of ω.

In this book $\mathbb{T}$ will always be a subset of the extended real line $\bar{\mathbb{R}}$: usually an interval of $\bar{\mathbb{R}}$ ("*continuous case*") or of $\bar{\mathbb{Z}}$ ("*discrete case*"), sometimes a dense countable set, for example. This is the situation in which the terminology originated : time, instants, and paths. But there also exist parts of the theory of processes where $\mathbb{T}$ is only a partially ordered set (in statistical mechanics, for example, $\mathbb{T}$ may be the family of subsets of a finite or countable set, partially ordered by inclusion) or even has no order structure at all (in some problems of ergodic theory $\mathbb{T}$ may be a group ; in problems concerning regularity of paths of Gaussian processes $\mathbb{T}$ is just a metric space). So this book, where the notion of time plays an essential role, gives a somewhat partial idea of the theory of processes.

Definition 1 calls for a number of remarks. 2

(a) Just as the notion of a random variable was related to a measurable space structure $(\Omega,\mathcal{F})$ and not to a probability space structure $(\Omega,\mathcal{F},\mathbb{P})$, the notion of a process does not really require a law $\mathbb{P}$, and from time to time we may speak of a process on some space, without emphasis on any particular law on it.

(b) We have defined a process as a family $(X_t)_{t \in \mathbb{T}}$ of r.v., i.e. a mapping of $\mathbb{T}$ into the set of E-valued random variables. A process can also be considered as a mapping $(t,\omega) \mapsto X_t(\omega)$ of $\mathbb{T} \times \Omega$ into E or as a mapping $\omega \mapsto (t \mapsto X_t(\omega))$ of Ω into the set of all possible paths. In the latter interpretation, the process appears as a random variable with values in the set of paths (a"random function"), but this notion is not complete from a mathematical point of view : it lacks a σ-field given on the set of all paths. We shall return to this.

The second point of view (a process is a function on $\mathbb{T} \times \Omega$) will be the most useful. We illustrate it by a definition :

3 DEFINITION. Suppose that $\mathbb{T}$ is given a σ-field $\mathcal{T}$. The process $(X_t)_{t \in \mathbb{T}}$ is said to be measurable if the mapping $(t,\omega) \mapsto X_t(\omega)$ is measurable on $\mathbb{T} \times \Omega$ with respect to the product σ-field $\mathcal{T} \times \mathcal{F}$.

In the discrete case ($\mathbb{T} \subset \bar{\mathbb{Z}}$), the σ-field $\mathcal{T}$ is that of all subsets of $\mathbb{T}$ and the notion is trivial : every process is measurable.

We continue the "remarks" on definition 1.

4 (c) A continuous time stochastic process is the kind of mathematical model one uses to describe a natural phenomenon whose evolution is governed by chance. Hence it is natural to wonder under which conditions two processes describe the same phenomenon. On the other hand, given a natural phenomenon, how can observations be used to construct a process which describes it ?

The classical answer to these questions is the following. Let us assume that at any finite system of instants $t_1, t_2,\ldots, t_n$, we can determine with arbitrary precision the state of the process. By performing a large number of independent experiments, it is then possible to estimate with arbitrary precision probabilities of the type

$$\mathbb{P}\{X_{t_1} \in A_1,\ldots, X_{t_n} \in A_n\} \quad (A_1,\ldots, A_n \in \mathcal{E}) \tag{4.1}$$

and in general observation can give nothing more. Hence the following definition expresses reasonably the fact that two processes (X_t) and (X'_t) represent the same natural phenomenon.

DEFINITION. We consider two stochastic processes with the same time set $\mathbb{T}$ and state space $(E,\mathcal{E})$:

$$(\Omega,\mathcal{F},\mathbb{P},(X_t)_{t \in \mathbb{T}}) \text{ and } (\Omega',\mathcal{F}',\mathbb{P}',(X'_t)_{t \in \mathbb{T}}).$$

The processes (X_t) and (X'_t) are called equivalent if :

$$\mathbb{P}\{X_{t_1} \in A_1, X_{t_2} \in A_2, \ldots, X_{t_n} \in A_n\} = \mathbb{P}'\{X'_{t_1} \in A_1, X'_{t_2} \in A_2, \ldots, X'_{t_n} \in A_n\}$$

for every finite system of instants $t_1, t_2, \ldots, t_n$ *and elements* $A_1, A_2, \ldots, A_n$ *of* $\mathcal{E}$.

Terminology is somewhat functuating : we often say that (X_t) and (X'_t) have the same *time law*, or simply the same *law*, or that they are *versions* of each other.

(d) However, the notion of a time law leads to criticism. On the one hand it 5
is *too* precise. For it is impossible in practice to determine a measure at any given instant. All that instruments can give are average results over small time intervals. In other words, we have no direct access to the r.v. X_t themselves, but only to r.v. of the form

$$\frac{1}{b-a}\int_a^b f(X_s)ds$$

where f is a function on the state space E (considering such integrals of course requires some measurability from the process). This leads to a notion of "almost-equivalence". We develop this topic in nos. 35-45.

(e) On the other hand, the time law notion is *insufficiently* precise, because it concerns only *finite* subsets of a set $\mathbf{T}$, which in general is uncountable. We take an example. On the probability space $\Omega = [0,1]$ with the Borel σ-field $\mathcal{F} = \mathcal{B}([0,1])$ and Lebesgue measure $\mathbb{P}$, we consider two real-valued processes (X_t) and (Y_t) defined as follows : $\mathbf{T} = [0,1] = \Omega$

(5.1)
$$X_t(\omega) = 0 \text{ for all } \omega \text{ and all } t$$
$$Y_t(\omega) = 0 \text{ for all } \omega \text{ and all } t \neq \omega,\; Y_t(t) = 1.$$

For *each* t, $Y_t = X_t$ a.s. but the set of ω such that $X_{\cdot}(\omega) = Y_{\cdot}(\omega)$ is empty. The two processes have the same time law but the first one has all its paths continuous while the paths of the second one are almost all discontinuous. In wouldn't be right to discard this example as artificial ; let us indeed give, for the expert, the following example : we consider a one-dimensional Brownian motion (B_t) starting from 0 and define

(5.2)
$$X_t(\omega) = 0 \text{ for all } \omega \text{ and all } t$$
$$Y_t(\omega) = 0 \text{ for all } \omega \text{ and all } t \text{ such that } B_t(\omega) \neq 0$$
$$= 1 \text{ for all } \omega \text{ and all } t \text{ such that } B_t(\omega) = 0.$$

The situation is the same as above and the process (Y_t) is by no means "artificial": it has been studied by Paul Lévy in a series of works which are considered masterpieces of probability theory (Lévy [1], Chapter VI).

We now give formal definitions of the notions we have just met. The first one is a little more precise than equivalence :

DEFINITION. *Let* $(X_t)_{t \in \mathbf{T}}$ *and* $(Y_t)_{t \in \mathbf{T}}$ *be two stochastic processes defined on the* 6
same probability space $(\Omega, \mathcal{F}, P)$ *with values in the same state space* $(E, \mathcal{E})$. *We say that*

$(Y_t)_{t \in \mathbb{T}}$ is a (standard) modification of $(X_t)_{t \in \mathbb{T}}$ if

$$X_t = Y_t \quad \text{a.s.}$$

for each $t \in \mathbb{T}$.

The second definition expresses the greatest possible precision from the probabilistic point of view : two indistinguishable processes really are "the same" process.

7 DEFINITION. In the notation of definition 6, the processes (X_t) and (Y_t) are called $\mathbb{P}$-indistinguishable (or simply indistinguishable) if for almost all $\omega \in \Omega$

$$X_t(\omega) = Y_t(\omega) \text{ for all } t.$$

For example, if two real-valued processes (X_t) and (Y_t) have right-continuous (or left-continuous) paths on $\mathbb{T} = \mathbb{R}$ and if, for each rational t, $X_t = Y_t$ a.s., then they are indistinguishable : the paths $X_.(\omega)$ and $Y_.(\omega)$ are indeed a.s. equal on the rationals and hence everywhere on $\mathbb{R}$.

8 The definition of indistinguishable processes can be expressed differently. A random set is a subset A of $\mathbb{T} \times \Omega$ whose indicator I_A, as a function of (t,ω), is a stochastic process (i.e. $\omega \mapsto I_A(t,\omega)$ is a r.v. for all t). The set A is said to be evanescent if the process I_A is indistinguishable from 0, which means also that the projection of A on Ω is contained in a $\mathbb{P}$-negligible set. Two processes (X_t) and (Y_t) then are indistinguishable if and only if the set $\{(t,\omega) : X_t(\omega) \neq Y_t(\omega)\}$ is evanescent.

Time laws : canonical process and construction

9 Among all the processes with a given time law, we try to distinguish some process defined unambiguously and naturally, using no information on the process other than its time law. Such a process is called canonical.

We consider a stochastic process $(\Omega,\mathcal{F},\mathbb{P},(X_t)_{t \in \mathbb{T}})$ with values in $(E,\mathcal{E})$. We denote by τ the mapping of Ω into $E^{\mathbb{T}}$ which associates with $\omega \in \Omega$ the point $(X_t(\omega))_{t\in\mathbb{T}}$ of $E^{\mathbb{T}}$, that is the path of ω. The mapping τ is measurable when $E^{\mathbb{T}}$ is given the product σ-field $\mathcal{E}^{\mathbb{T}}$ (see I.8) ; hence we can consider the image law $\tau(\mathbb{P})$ on the space $(E^{\mathbb{T}},\mathcal{E}^{\mathbb{T}})$. We denote by Y_t the coordinate mapping of index t on $E^{\mathbb{T}}$. The processes $(\Omega,\mathcal{F},\mathbb{P},(X_t)_{t \in \mathbb{T}})$ and $(E^{\mathbb{T}},\mathcal{E}^{\mathbb{T}},\tau(\mathbb{P}), (Y_t)_{t \in \mathbb{T}})$ are then equivalent (by the very definition of image laws) and we can set the definition :

DEFINITION. In the above notation, the process

$$(E^{\mathbb{T}},\mathcal{E}^{\mathbb{T}},\tau(\mathbb{P}), (Y_t)_{t \in \mathbb{T}})$$

is called the canonical process associated with (or equivalent to) (X_t).

Two processes (X_t) and (X'_t) are equivalent if and only if they are associated with the same canonical process.

This canonical process is hardly ever used directly when the time set $\mathbb{T}$ is uncountable : the σ-field $\mathcal{E}^{\mathbb{T}}$ contains just events which depend only on countably many variables Y_t, whereas the most interesting properties of the process (continuity of paths, for example) involve all these random variables. The canonical process is mainly useful as a step in the construction of more complicated processes.

> We must insist on the fact that the "canonical" character depends on the available information on the process (X_t). In the absence of any information other than the time law, everybody will be satisfied with the above canonical process. But if it is known, for example, that the process (X_t) has a version with continuous paths (under some topology on $\mathbb{T}$), then it would by silly to use it. The set $E^{\mathbb{T}}$ of all mappings of $\mathbb{T}$ into E will be replaced by that of all <u>continuous</u> mappings of $\mathbb{T}$ into E, onto which the measure will be carried by the same procedure as above, thus defining a canonical <u>continuous</u> process.

The notion of a canonical process leads to a simple - but hardly satisfying - solution to the problem of constructing stochastic processes. 10

We return to the situation described in no. 4 : we have observed some "random phenomenon" which we wish to represent by means of a process. Since it can only be defined to within an equivalence, the choice that offers itself to the mind is that of the canonical process. Hence we use the measurable space $(E^{\mathbb{T}},\mathcal{E}^{\mathbb{T}})$ and the coordinate mappings $(Y_t)_{t \in \mathbb{T}}$. It remains to construct a probability law $\mathbb{P}$ on this space such that

$$\mathbb{P}\{Y_{t_1} \in A_1,\dots, Y_{t_n} \in A_n\} = \Phi(t_1,\dots, t_n ; A_1,\dots, A_n)$$

for every finite subset $\underline{u} = \{t_1, t_2,\dots, t_n\}$ of $\mathbb{T}$ and every finite family $A_1, A_2,\dots, A_n$ of measurable subsets of E, the functions Φ being given by observation. For the construction to be possible, it is necessary that the set function

$$A_1 \times A_2 \times \dots \times A_n \mapsto \Phi(t_1, t_2,\dots, t_n ; A_1, A_2,\dots, A_n)$$

be extendedable to a probability law $\mathbb{P}_{\underline{u}}$ on $(E^{\underline{u}}, \mathcal{E}^{\underline{u}})$, probability law which moreover is uniquely determined by Φ (by Theorem I.20, applied to the set of finite unions of subsets of $E^{\underline{u}}$ of the form $A_1 \times A_2 \times \dots \times A_n$). On the other hand it is necessary that

$$\pi_{\underline{u}\underline{v}}(\mathbb{P}_{\underline{v}}) = \mathbb{P}_{\underline{u}}$$

for every pair of finite subsets $\underline{u}$, $\underline{v}$ of T such that $\underline{u} \subset \underline{v}$, where $\pi_{\underline{u}\underline{v}}$ denotes the projection of $E^{\underline{v}}$ onto $E^{\underline{u}}$. We recognize here the definition of an inverse system of probability laws (III.52) and the possibility of constructing the law $\mathbb{P}$ appears to be equivalent to the existence of an inverse limit for the inverse system $(\mathbb{P}_{\underline{u}})$. Theorem III.52 then gives a simple condition that implies the existence of $\mathbb{P}$.

Adapted and progressive processes

<u>We henceforth assume that the time set</u> $\mathbb{T}$ <u>is the closed positive half-line</u> $\mathbb{R}_+$. We leave to the reader all trivial extensions to other time sets, except for a few remarks on more delicate points. In the numbers which follow we introduce some terminology which will be used throughout this book, but we postpone until paragraph 3 a detailed study of it.

Let $(\Omega,\mathcal{F})$ be a measurable space and let $(\mathcal{F}_t)_{t \in \mathbb{R}}$ be a family of sub-σ-fields of $\mathcal{F}$ 11 such that $\mathcal{F}_s \subset \mathcal{F}_t$ for $s \leq t$. We shall say that $(\mathcal{F}_t)$ is an increasing family of σ-fields on $(\Omega,\mathcal{F})$ or a filtration of $(\Omega,\mathcal{F})$: $\mathcal{F}_t$ is called the σ-field of events prior to t. We define

$$(11.1) \qquad \mathcal{F}_{t+} = \bigcap_{s>t} \mathcal{F}_s , \quad \mathcal{F}_{t-} = \bigvee_{s<t} \mathcal{F}_s \quad (t > 0).$$

The family $(\mathcal{F}_t)$ is said to be right-continuous if $\mathcal{F}_t = \mathcal{F}_{t+}$ for all t. For example the family $(\mathcal{F}_{t+})_{t \in \mathbb{R}}$ is right-continuous for every family $(\mathcal{F}_t)$. The analogous concept on the left-hand side does not seem to be of interest.

When the time set is $\mathbb{N}$, definitions (11.1) still have a meaning : $\mathcal{F}_{n+}$ and $\mathcal{F}_{n-}$ must be interpreted as $\mathcal{F}_{n+1}$ and $\mathcal{F}_{n-1}$. It turns out that the latter analogy between $\mathcal{F}_{n-1}$ and $\mathcal{F}_{t-}$ is interesting, while the former isn't.

12 DEFINITION. Let $(X_t)_{t \in \mathbb{R}_+}$ be a process defined on a measurable space $(\Omega,\mathcal{F})$ and let $(\mathcal{F}_t)_{t \geq 0}$ be a filtration. The process (X_t) is said to be adapted to $(\mathcal{F}_t)$ if X_t is $\mathcal{F}_t$-measurable for every $t \in \mathbb{R}_+$.

EXAMPLE. Every process (X_t) is adapted to the family of σ-fields $\mathcal{F}_t = \sigma(X_s, s \leq t)$, which is often called the natural family of this process.

13 The above definitions have the following intuitive meaning : if we interpret the parameter t as time and each event as a physical phenomenon, the sub-σ-field $\mathcal{F}_t$ consists of the events which represent phenomena prior to the instant t. The $\mathcal{F}_t$-measurable random variables are therefore those which depend only on the evolution of the universe prior to t. Obviously the two processes representing "the result of the horse races of last Sunday" and "the result of the horse races of next Sunday" play very different roles in the universe of the observer, because the first one is adapted and the second is not. And yet, the first process may possibly be (in first approximation) stationary in time, so that both processes have the same law. Hence it is really the introduction of a filtration which expresses that the parameter t is a time, that the future is uncertain whereas the past is knowable (at least for an ideal observer). This fundamental idea is due to Doob. From the mathematical point of view, the presence of a filtration in the hypotheses is not a restriction, for it is permissible to take $\mathcal{F}_t = \mathcal{F}$ for all t. This choice corresponds to the deterministic world view of the XIX th century, where the ideal observer may predict at any time, through the integration of a complicated differential system, all the future evolution of the universe. If there has ever been any "real" intervention of chance, it has taken place at the initial instant, and causality has left no room for it thereafter. This, however, does not quite prevent probabilities from occuring in this description of the universe, because of the imprecise nature of our measurements. These comments may explain why the filtration $\mathcal{F}_t = \mathcal{F}$ is often called the deterministic filtration.

14 On a space $(\Omega,\mathcal{F})$ filtered by a family $(\mathcal{F}_t)_{t \geq 0}$, the notion of a measurable process may be made precise as follows :

DEFINTION. Let $(\Omega,\mathcal{F})$ be a measurable space and let $(\mathcal{F}_t)_{t\geq 0}$ be a filtration of it. Let $(X_t)_{t\geq 0}$ be a process defined on this space with values in $(E,\mathcal{E})$; we say that (X_t) is progressively measurable or progressive with respect to the family $(\mathcal{F}_t)$ if for every $t \in \mathbb{R}_+$ the mapping $(s,\omega) \mapsto X_s(\omega)$ of $[0,t] \times \Omega$ into $(E,\mathcal{E})$ is measurable with respect to the σ-field $\mathcal{B}([0,t]) \times \mathcal{F}_t$.

A progressive process is obviously adapted.

How can we check that a given process is progressive ? One may often reduce to apparently weaker conditions :

(a) if (X_t) is adapted and for every $\varepsilon > 0$ is progressive with respect to the family $(\mathcal{F}_{t+\varepsilon})_{t\geq 0}$, then it is progressive with respect to $(\mathcal{F}_t)$;

(b) if the adaptation hypothesis is omitted, one can still assert that (X_t) is progressive with respect to the family $(\mathcal{F}_{t+})$.

Both cases are treated in the same way : we have to study sets of the form $\{(t,\omega) : X_t(\omega) \in A\}$, $A \in \mathcal{E}$, and can hence reduce to real-valued processes. Then for $s \leq t$ we have

$$X_s = \lim_{\varepsilon\to 0} X_s I_{[0,t-\varepsilon[}(s) + X_t I_{\{t\}}(s) \tag{14.1}$$

and the right-hand side is, for all $\varepsilon > 0$, a measurable function with respect to $\mathcal{B}([0,t]) \times \mathcal{F}_t$ (or $\mathcal{F}_{t+}$ if X_t is only $\mathcal{F}_{t+}$-measurable).

If the intervals $[0,t]$ in definition 14 are replaced by intervals $[0,t[$, one gets only progressivity with respect to the family $(\mathcal{F}_{t+})$.

Here is the easiest example of a progressive process. It will be studied more deeply in paragraph 3.

THEOREM. Let E be a metrizable space and (X_t) a process with values in E, adapted 15
to $(\mathcal{F}_t)$ and with right-continuous paths. Then (X_t) is progressive with respect to $(\mathcal{F}_t)$. The same conclusion holds for a process with left-continuous paths.

Proof : For every $n \in \mathbb{N}$ we define

$$X_t^n = X_{(k+1)2^{-n}} \quad \text{if } t \in [k2^{-n}, (k+1)2^{-n}[.$$

This process is obviously progressive with respect to the family $(\mathcal{F}_{t+\varepsilon})$ provided $\varepsilon > 2^{-n}$. Hence the process X_t, equal to $\lim_n X_t^n$ by right-continuity, is progressive with respect to each family $(\mathcal{F}_{t+\varepsilon})$. Since it is adapted, we conclude by 14 (a). The argument is analogous for left-continuity.

2. REGULARITY OF PATHS

A path governed by chance - that of a particle getting hit from all sides, for example - has no reason for being very regular. The problems that arise naturally about paths of such real-valued processes are of the following kind :

- are the paths measurable ? locally integrable ?

(or : does the process have a modification with these properties ?) ;

- are the paths locally bounded ? are the quantities of the type $\sup_{t\in I} |X_t|$ random variables when I is an interval (and hence a non-countable set) ? does there exist a method for determining their law ?

- are the paths continuous at a point ? continuous on an interval ?

Continuity itself appears in this context as a strong property, since typical paths of "nice" processes may have jumps (with limits on both sides).

This paragraph is divided into three parts. First, the study of paths along a countable dense set (nos. 17-23), leading to the theory of separability (nos. 24-30). We immediately warn the reader that separability will not be used later and can be omitted without inconvenience. Next, the direct study of a measurable process on the whole of $\mathbb{R}_+$, using the theory of analytic sets (nos. 31-34). Finally, the study of paths of processes "up to negligible sets" (nos. 35-45).

It is important that the reader should keep this plan in mind, since the same properties are studied three times from three different points of view : see for examples Theorems 18-19, then 34, then 46. Similarly, 17, 33 and 38.

16 NOTATION

Throughout this paragraph, the time set of the processes is $\mathbb{R}_+$ unless otherwise mentioned. They are defined on a probability space $(\Omega,\mathcal{F},\mathbb{P})$ provided with a filtration $(\mathcal{F}_t)_{t \geq 0}$. We use the following abbreviation : r.c. for "right-continuous", l.c. for "left continuous", r.c.l.l. (the most important one) for "right-continuous on $[0,+\infty[$ with finite left hand limits on $]0,+\infty[$"(1). The reader will be able to conjecture the meaning of l.c.r.l. and l.l.l.r. in a similar way.

Processes on a countable dense set

17 Here D is a countable dense subset of $\mathbb{R}_+$, $(X_s)_{s \in D}$ is an adapted real-valued (i.e. X_s is $\mathcal{F}_s$-measurable for every $s \in D$).

THEOREM. <u>For all</u> $t \geq 0$ <u>we set</u> (2)

(17.1) $$\bar{Y}_t^+(\omega) = \limsup_{s \in D, s \downarrow t} X_s(\omega),\quad \underline{Y}_t^+(\omega) = \liminf_{s \in D, s \downarrow t} X_s(\omega).$$

<u>Then the two processes</u> $(\bar{Y}_t^+)$ <u>and</u> $(\underline{Y}_t^+)$ <u>are progressive relative to the family</u> $(\mathcal{F}_{t+})$. <u>So are the processes defined on</u> $]0,\infty[$ <u>by</u>

(17.2) $$\bar{Z}_t^-(\omega) = \limsup_{s \in D, s \uparrow\uparrow t} X_s(\omega),\quad \underline{Z}_t^-(\omega) = \liminf_{s \in D, s \uparrow\uparrow t} X_s(\omega)$$

---

(1) The French abbreviation is càdlàg (continue à droite limites à gauche), which sounds perfectly English to any reader of Gulliver's Travels, and has found its way into many papers in english.

(2) We emphasize that $\lim_{s \to t}$ or $\lim_{s \uparrow t}$ are limits with t included, so that if $\lim_{s \to t} f(s)$ exists, f must be continuous at t. On the other hand, $\lim_{s \to t, s \neq t}$ or $\lim_{s \uparrow\uparrow t}$ exclude t. This conforms to the "French" (bourbakist) use, but isn't standard in English texts.

relative to the family $(\mathcal{F}_t)$.

Proof : For every integer n we define a process (Y_t^n) as follows :

$$\text{if } t \in [k2^{-n}, (k+1)2^{-n}[,\ Y_t^n = \sup_{s \in D_t} X_s, \text{ where } D_t = D \cap]t, (k+1)2^{-n}[.$$

This process is adapted to the family $(\mathcal{F}_{t+\varepsilon})$ for all $\varepsilon > 2^{-n}$, it is right-continuous and hence progressive relative to this family (15). Therefore so is the process

$$Y_t = \limsup_{s \in D, s \uparrow\uparrow t} X_s = \lim_n Y_t^n.$$

It follows from 14 that (Y_t) is progressive with respect to $(\mathcal{F}_{t+})$. To deal with $(\bar{Y}_t^+)$, we simply note that

$$\bar{Y}_t^+ = Y_t I_{\{t \notin D\}} + (Y_t \vee X_t) I_{\{t \in D\}}.$$

The argument is similar for the other processes (17.1) and (17.2).

The following statement is the first in a series which will be pursued throughout this paragraph. We keep the notation of 17.

THEOREM. (a) The set W (resp. W') of $\omega \in \Omega$ such that the path $X(\omega)$ is the restriction to D of a right-continuous (resp. r.c.l.l.) mapping on $\mathbb{R}_+$ is the complement of an $\mathcal{F}$-analytic set (hence it belongs to the universal completion σ-field of $\mathcal{F}$). This result extends to processes with values in a cosouslin metrizable space E. 18

(b) In the real case, or more generally for processes with values in a Polish space E, it can even be affirmed that W' belongs to $\mathcal{F}$.

Proof : We begin (a). We imbed the separable metrizable space E in the cube $I = \bar{\mathbb{R}}^{\mathbb{N}}$ (or simply in $\bar{\mathbb{R}}$ if $E = \mathbb{R}$) and denote by J the compact metrizable space obtained by adjoining an isolated point α to I. We write

$$X_{t+}(\omega) = \lim_{s \in D, s \downarrow t} X_s(\omega) \text{ if this limit exists in I}$$

$$= \alpha \text{ otherwise}$$

and similarly $X_{t-}(\omega) = \lim_{s \in D, s \uparrow\uparrow t} X_s(\omega)$ if this limit exists, $= \alpha$ otherwise.

For $t \in D$, the existence of $X_{t+}(\omega)$ implies $X_{t+}(\omega) = X_t(\omega)$.(1) The mappings $(t,\omega) \mapsto X_{t+}(\omega)$, $X_{t-}(\omega)$ are $\mathcal{B}(\mathbb{R}_+) \times \mathcal{F}$-measurable. Since $I = \bar{\mathbb{R}}^{\mathbb{N}}$, we can indeed immediately reduce to the real case and then

$$X_{t+}(\omega) = \bar{Y}_t^+(\omega) \text{ if } \bar{Y}_t^+(\omega) = \underline{Y}_t^+(\omega),\ X_{t+}(\omega) = \alpha \text{ otherwise}$$

and we apply 17 ; similarly for X_{t-} using $\bar{Z}^-$, $\underline{Z}^-$. We denote by A the set $J \setminus E$; since E is cosouslin, A is analytic in J (III.19) and the set

$$H = \{(t,\omega) : X_{t+}(\omega) \in A\} \text{ (resp. } H' = \{(t,\omega) : X_{t+}(\omega) \in A \text{ or } X_{t-}(\omega) \in A\})$$

is analytic, being the inverse image of an analytic set under a measurable mapping (III.11). We conclude by noting that the complement of W (resp. W') is the projection of H (resp. H') onto Ω and applying III.13.

It remains to show that W' is $\mathcal{F}$-measurable if E is Polish. We shall see later (no. 23) a similar result proved quite differently.

(1) Since X_{t+} is a right limit at t with t included.

We give E a metric d, under which E is complete, and set $d(\alpha,E) = +\infty$. For $\varepsilon > 0$ we define the following functions inductively :

$$T_0^{\varepsilon}(\omega) = 0$$

$$Z_0^{\varepsilon}(\omega) = \lim_{s\in D, s\downarrow 0} X_t(\omega) \text{ if this limit exists}$$

$$= \alpha \text{ otherwise}$$

then

$$T_{n+1}^{\varepsilon}(\omega) = \inf\{t \in D,\ t > T_n(\omega),\ d(X_t(\omega), Z_n(\omega)) > \varepsilon\} \quad (\inf \emptyset = +\infty)$$

$$Z_{n+1}^{\varepsilon}(\omega) = \lim_{s\in D, s\downarrow T_{n+1}^{\varepsilon}(\omega)} X_s(\omega) \text{ if } T_{n+1}^{\varepsilon}(\omega) < \infty \text{ and this limit exists,}$$

$$= \alpha \text{ otherwise.}$$

It is easy to verify that the functions T_n^{ε}, Z_n^{ε} on E^D are $\mathfrak{F}$-measurable. The statement then follows from the following lemma.

LEMMA.

$$W' = \{\omega \in \Omega : \forall k \in \mathbb{N},\ \lim_n T_n^{2^{-k}}(\omega) = +\infty\}.$$

Proof : (a) If $\omega \in W'$, ω is the restriction to D of a right-continuous mapping on $\mathbb{R}_+$ into E. It follows by right-continuity that the limits in the preceding definition always exist and that for all ε and all n such that $T_n^{\varepsilon}(\omega) < \infty$

$$T_{n+1}^{\varepsilon}(\omega) > T_n^{\varepsilon}(\omega),\ d(Z_n^{\varepsilon}(\omega), Z_{n+1}^{\varepsilon}(\omega)) \geq \varepsilon \text{ if } T_{n+1}^{\varepsilon}(\omega) < \infty.$$

The oscillation of ω on the interval $[T_n^{\varepsilon}(\omega), Z_{n+1}^{\varepsilon}(\omega)]$ is therefore at least ε if $T_{n+1}^{\varepsilon}(\omega) < \infty$, and the existence of left-hand limits therefore prevents the $T_n^{\varepsilon}(\omega)$ from accumulating at a finite distance. Consequently $\lim_n T_n^{\varepsilon}(\omega) = +\infty$ for all $\varepsilon > 0$.

(b) Conversely, suppose that $T_n^{\varepsilon}(\omega) \mapsto +\infty$. We define a r.c.l.l. mapping f_ε of D into E by writing

$$f_\varepsilon(t) = Z_n^{\varepsilon}(\omega) \text{ for } t \in D \cap [T_n^{\varepsilon}(\omega), T_{n+1}^{\varepsilon}(\omega)[.$$

Then $d(X_t(\omega), f_\varepsilon(t)) \leq 2\varepsilon$ for all $t \in D$. If the above property is satisfied for values of ε tending to 0 - for example $\varepsilon = 2^{-k}$ - we see that $X_.(\omega)$ is the uniform limit on D of a sequence of r.c.l.l. mappings on $\mathbb{R}_+$. It follows immediately that it can be extended to a r.c.l.l. mapping on $\mathbb{R}_+$.

19 REMARKS. (a) Theorem 18 can be put into a "canonical" form. We consider the set W (resp. W') of all right-continuous (resp. r.c.l.l.) mappings of $\mathbb{R}_+$ into a cosouslin metrizable space E and give it the σ-field generated by the coordinate mappings. The mapping which associates with each $w \in W$ (resp. W') its restriction to D is a measurable isomorphism of W (resp. W') into $\Omega = E^D$ with its Borel σ-field ; Ω is cosouslin and Polish if E is Polish. Applying 18 to the process $(X_t)_{t \in D}$ consisting of the coordinate mappings on Ω, it follows that W is the complement of an analytic set in Ω , and W' a Borel subset of Ω if E is Polish. Hence the measurable space W is cosouslin and the measurable space W' is Lusin if E is Polish. The proof also indi-

cates a cosouslin (resp. Lusin) topology on W (resp. W'), that of pointwise convergence on D, but this topology is uninteresting in general, since it involves an arbitrary choice of a countable dense set D, while the measure theoretic statement is instrinsic.

When E is Polish, there exists an interesting topology on W' under which W' is Polish : the Skorokhod topology. See for example Maisonneuve [1].

(b) We adjoin to E an isolated point denoted by ∂ and denote by Ω the set of all right-continuous mappings ω of $\mathbb{R}_+$ into $E \cup \{\partial\}$, which keep the value ∂ from the first instant they assume it, so that the set $\{t : \omega(t) = \partial\}$ is a closed halfline (possibly empty) $[\zeta(\omega),+\infty[$. It is easy to see that the lifetime ζ is a measurable function relative to the σ-field $\mathcal{F}^0$ on Ω generated by the coordinate mappings if E is cosouslin, and that Ω is a cosouslin space under the topology of pointwise convergence on D. On the other hand, if E is Polish, the space Ω' of elements ω of Ω with a left limit in E at every point of the interval $]0,\zeta(\omega)[$ (but not necessarily at the instant $\zeta(\omega)$ itself) is Lusin under the topology of pointwise convergence on D. The idea is the same as in the proof of 13, one just has to write $\lim_n T_n^\varepsilon \geq \zeta$ instead of $\lim_n T_n^\varepsilon = +\infty$.

Upcrossings and downcrossings

We now indicate a variant of the preceding results. We no longer require right-continuity and the existence of left-hand limits, but rather existence of right-hand and left-hand limits. The interest of the results themselves is slight, but the numbers of upcrossings and downcrossings are important in martingale theory.

Let f be a mapping of $\mathbb{R}_+$ into a Hausdorff space E. We say that f is free of oscillatory discontinuities [(1)] if the right-hand limit 20

$$f(t+) \lim_{s \downarrow\downarrow t} f(s)$$

exists in E at every point t of $\mathbb{R}_+$ and the left-hand limit

$$f(t-) = \lim_{s \uparrow\uparrow t} f(s)$$

also exists in E at every point of $\mathbb{R}_+ \setminus \{0\}$ (it does not necessarily exist at infinity). We begin by considering extended real-valued functions and giving (following Doob) a simple criterion of freedom from oscillatory discontinuities in $\bar{\mathbb{R}}$.

Let f be a mapping of $\mathbb{R}_+$ into $\bar{\mathbb{R}}$. We denote by a, b two finite real numbers such that a < b and by $\underset{\sim}{u}$ a finite subset of $\mathbb{R}_+$ whose elements are $s_1, s_2, \ldots, s_n$ (arranged by order of magnitude). We define inductively the instants $t_1, \ldots, t_n \in \underset{\sim}{u}$ as follows: 21

t_1 is the first of the elements s_i of $\underset{\sim}{u}$ such that $f(s_i) < a$, or s_n if no such element exists ;

t_k is, for every even (resp. odd) integer lying between 1 and n, the first of the elements s_i of $\underset{\sim}{u}$ such that $s_i > t_{k-1}$ and $f(s_i) > b$ (resp. $f(s_i) < a$). If no such element exists, we write $t_k = s_n$.

---

(1) Bourbaki calls this a "regulated function". Our abbreviation should be "r.l.l.l".

We consider the last even integer 2k such that

$$f(t_{2k-1}) < a,\ f(t_{2k}) > b\ ;$$

if no such integer exists we write k = 0. The intervals

$$(t_1,t_2),\ (t_3,t_4),\ldots,\ (t_{2k-1},t_{2k})$$

of $\underset{\sim}{u}$ represent periods of time during which the function f goes upward, from below a to above b, whereas the intermediate intervals represent downward periods. The number k is called the number of upcrossings by f (considered on $\underset{\sim}{u}$) of the interval [a,b] and is denoted by

(21.1) $$U(f;\underset{\sim}{u};[a,b]).$$

We define similarly the number of downcrossings of f (considered on u) on the interval [a,b]:

(21.2) $$D(f;\underset{\sim}{u};[a,b]) = U(-f,\underset{\sim}{u},[-b,-a]).$$

We can also define the upcrossings and downcrossings of an interval of the form]a,b[, replacing strict inequalities by loose inequalities in the definition of the instants t_i (1).

Now let S be any subset of $\mathbb{R}_+$. We write :

(21.3) $$U(f;S;[a,b]) = \sup_{\substack{\underset{\sim}{u}\text{ finite}\\ \underset{\sim}{u}\subset S}} U(f;\underset{\sim}{u};[a,b]).$$

Definition (21.2) can be similarly extended.

The principal interest of these numbers arises from the following theorem :

22 THEOREM. Let f be a function on $\mathbb{R}_+$ with values in $\bar{\mathbb{R}}$. For f to be free of oscillatory discontinuities, it is necessary and sufficient that

(22.1) $$U(f;I;[a,b]) < +\infty$$

for every pair of rational numbers a, b such that a < b and every compact interval I of $\mathbb{R}_+$.

Proof : Suppose that there exists a point t where the function f has an oscillatory discontinuity, for example, where it has no left-hand limit. Then we can find a sequence of points t_n increasing to t such that

$$\liminf_{\substack{n\to\infty\\ n\text{ odd}}} f(t_n) = c > d = \limsup_{\substack{n\to\infty\\ n\text{ even}}} f(t_n).$$

we then choose a sufficiently large interval I and two rational numbers a and b such that d < a < b < c. It is immediately verified, removing finite subsets from the set of points t_n, that $U(f;I;[a,b]) = +\infty$.

The converse follows from a property which the reader can prove easily : if

---

(1) The numbers $U(f;\underset{\sim}{u};[a,b])$, $D(f;\underset{\sim}{u};[a,b])$ have the advantage of defining lower semi-continuous functions of f for pointwise convergence. This property extends to the number of upcrossings or downcrossings on any set S.

r, s, t are three instants such that $r < s < t$, then :

$$U(f;[r,t];[a,b]) \leq U(f;[r,s];[a,b]) + U(f;[s,t];[a,b]) + 1.$$

Let α and β be the end-points of I. Suppose that the function f has no oscillatory discontinuities ; then we can associate with each point $t \in I$ an open interval I_t containing t, such that the oscillation of f on each one of the intervals $I_t \cap]t,\beta]$, $[\alpha,t[\cap I_t$ is strictly less than b - a. We can cover the interval I with a finite number of intervals $I_{t_1}, I_{t_2}, \ldots, I_{t_k}$. We arrange by order of magnitude the points α and β, the points $t_1, t_2, \ldots, t_k$ and the end-points of the intervals $I_{t_1}, I_{t_2}, \ldots, I_{t_k}$; we thus get a finite set of points : $\alpha = s_0 < s_1 < \ldots < s_n = \beta$, such that the oscillation of f on each of the intervals $]s_i, s_{i+1}[$ is no greater than b - a. Then we have $U(f;]s_i, s_{i+1}[,[a,b]) = 0$ and consequently also $U(f;[s_i, s_{i+1}],[a,b]) \leq 1$. The inequality quoted above then gives

$$U(f: I;[a,b]) \leq 2n - 1$$

and the converse is established.

REMARKS. (a) The above statement concerns $\bar{\mathbb{R}}$. One can also express, using numbers of upcrossings, whether a finite function f on $\mathbb{R}_+$ has finite right-hand and left-hand limits. For a finite function with finite right-hand and left-hand limits is bounded in the neighbourhood of every point and hence bounded on every compact interval I, so that for every rational a

(22.2) $\quad \lim_n U(f;I;[a,a+n]) = 0 = \lim_n U(f;I;[a-n,a])$.

Whereas conversely, if f is not bounded above for example, we can find some a such that the left-hand side of (22.2) is ≥ 1 for all n.

Here is the application to stochastic processes.

THEOREM. Let E be an LCC space and let $(X_t)_{t \in D}$ be a process with values in E, defined on $(\Omega,\mathcal{F},\mathbb{P})$ with time set a countable dense set D. The set of all $\omega \in \Omega$ such that the path $X_{\cdot}(\omega)$ on D can be extended to a mapping of $\mathbb{R}_+$ into E without oscillatory discontinuities is $\mathcal{F}$-measurable. 23

Proof : We may assume that E is the complement of a point x_0 in a compact metric space F, whose distance is denoted by d. Let $(x_n)_{n \geq 1}$ be a sequence dense in E. We write $h_n(x) = d(x_n,x)$ for $n \geq 1$ (so that the sequence (h_n) of continuous functions separates the points) and $h_0(x) = 1/d(x_0,x)$. We want to express that each one of the real processes $(h_n \circ X_t)_{n \geq 1}$ has right-hand and left-hand limits along D and that the process $(h_0 \circ X_t)$ has finite right-hand and left-hand limits along D. This follows immediately using the numbers of upcrossings of paths, considered on D.

REMARKS. (a) The result extends to the case of a Polish space E, since every Polish space E can be considered (III.17) as a $\mathcal{G}_\delta$ in some compact metric space and hence as an intersection of LCC spaces E_n. We then write down the preceding conditions for each of the E_n. If E were cosouslin (in particular Lusin), the set in the statement would be the complement of an $\mathcal{F}$-analytic set : we leave this aside.

(b) We have been concerned here with r.c.l.l. or r.l.l.l. mappings, but we might consider continuous mappings analogously. The method would be more classical: To express that a mapping of D into a Polish space E can be extended to a continuous mapping of $\mathbb{R}_+$ into E, one just writes for every integer n the condition for uniform continuity on $D \cap [0,n]$.

Choice of the countable set : separability

We emphasize again that the results of nos. 24 to 30 will not be used elsewhere and can therefore be omitted.

24 Our problem now is the following : how can we recognize whether a given process (X_t) admits a modification (Y_t) with nice properties - for example, a modification with r.c.l.l. paths or (which is more difficult) a modification with bounded paths. This is a quite natural problem, and we shall in fact study later on another problem of the same kind, relative to "almost-modifications". However, it is sometimes forbidden to modify a given process. Recall the example, already considered in no. 5, of the process

(24.1) $$X_t = 0 \text{ if } B_t \neq 0,\ X_t = 1 \text{ if } B_t = 0,$$

where (B_t) is one dimensional brownian motion, with continuous paths. If we are just looking for a modification with regular paths, we may spare our time and simply take the modification $Y_t = 0$. Here the theory of separability would destroy the structure of the process.

The theory of separability was developed by Doob for continuous time processes $(X_t)_{t \in \mathbb{R}_+}$. It extends without difficulty to processes whose time set is a topological space with countable base. Instead of this, we study processes indexed by $\mathbb{R}_+$, but under the right topology (1) on $\bar{\mathbb{R}}$, which hasn't a countable base. This extension is due to Chung (Chung-Doob [2]). On the other hand, the theory can be extended to processes with values in a compact metrizable space, whereas we only consider processes with values in $\bar{\mathbb{R}}$ (beware, the distinction between $\bar{\mathbb{R}}$ and $\mathbb{R}$ is important here).

25 The definition may be clearer in its most general form.

DEFINITION. *Let* f *be a mapping of a topological space* T *into a topological space* E *and let* D *be a dense set in* T. *We say that* f *is* D-*separable if the set of points* $(t,f(t))$, $t \in D$, *is dense in the graph of* f *(for the product topology on* $T \times E$).

Henceforth we take $T = \mathbb{R}_+$ with the right topology and $E = \bar{\mathbb{R}}$. On the other hand, D will be countable. We then say that f is *right* D-*separable* (D-*separable* if the ordinary topology of $\mathbb{R}_+$ is used.)

(1) Recall that the neighbourhoods of $x \in \mathbb{R}_+$ for the right topology are the sets containing an interval $[x, x+\varepsilon[$, $\varepsilon > 0$, so that a *left*-closed interval $[a,b[$ is closed under the *right* topology !

DEFINITION. Let $(X_t)_{t \in \mathbb{R}_+}$ be a process with values in $\bar{\mathbb{R}}$, defined on a probability 26
space $(\Omega, \mathfrak{F}, \mathbb{P})$. (X_t) is called right separable if there exists a countable dense set D such that, for almost all $\omega \in \Omega$, the path $X_{.}(\omega)$ is right D-separable.

If (X_t) is right separable, we may solve easily the problem of no. 24 : if the paths are bounded on D, they are bounded everywhere, etc... The following lemma is a modification due to Chung of a result of Doob (*Stochastic Processes*, pp. 56-57).

THEOREM. Let $(X_t)_{t \in \mathbb{R}_+}$ be a process with values in $\bar{\mathbb{R}}$. There exists a countable dense 27
set D with the following property : for every closed set F of $\bar{\mathbb{R}}$ and every set $I \subset \mathbb{R}_+$ open under the right topology,

(27.1) $\qquad \mathbb{P}\{X_t \in F$ for all $t \in D \cap I,\ X_u \in F\} = 0$ for every $u \in I$ and, for every countable set S

(27.2) $\qquad \mathbb{P}\{X_t \in F$ for all $t \in D \cap I\} \leq \mathbb{P}\{X_t \in F$ for all $t \in S \cap I\}$.

Proof : We leave to the reader the equivalence of (27.1) and (27.2), which is easy. We choose a countable set $\mathcal{H}$ of closed subsets of $\bar{\mathbb{R}}$, such that every closed set is the intersection of a descreasing sequence of elements of $\mathcal{H}$, and a countable set $\mathcal{G}$ of open subsets of $\mathbb{R}_+$ with the ordinary topology, such that every (ordinary) open set of $\mathbb{R}_+$ is the union of an increasing sequence of elements of $\mathcal{G}$. For every pair (I,F), $I \in \mathcal{G}$, $F \in \mathcal{H}$, we choose a countable set $\Delta(I,F)$ dense in I such that the probability

$$\mathbb{P}\{X_t \in F \text{ for all } t \in S \cap I\} \quad (S \text{ countable})$$

is minimal for $S = \Delta(I,F)$. We set $\Delta(F) = \cup\, \Delta(I,F)$ for $I \in \mathcal{G}$. Then for every ordinary open set I and every countable set S we have

(27.3) $\qquad \mathbb{P}\{X_t \in F$ for all $t \in \Delta(F) \cap I\} \leq \mathbb{P}\{X_t \in F$ for all $t \in S \cap I\}$.

Always keeping $F \in \mathcal{H}$ fixed, we consider for r rational > 0 the increasing function on $[0,r[$

$$h_r(t) = \inf_S \mathbb{P}\{X_u \in F \text{ for all } u \in S \cap [t,r[\} \quad (S \text{ countable})$$

which we compare to

$$k_r(t) = \mathbb{P}\{X_u \in F \text{ for all } u \in \Delta(F) \cap [t,r[\}.$$

We have, by the choice of $\Delta(F)$, $h_r(t+) = k_r(t+)$ for all t and hence h_r and k_r differ only on a countable set N_r. If we enlarge $\Delta(F)$ by replacing it - without changing the notation - by $\Delta(F) \cup (\bigcup_r N_r)$, we have, for every rational r and every $t \in [0,r]$, $h_r(t) = k_r(t)$. But then the same result will hold for all real r on passing to the limit. Thus, for every interval $[t,r[$

(27.4) $\qquad \mathbb{P}\{X_u \in F$ for all $u \in \Delta(F) \cap [t,r[\,,\ X_t \notin F\} = 0$.

Now let I be an open set under the right topology : I is a countable union of disjoint intervals of the form $]t_i, r_i[$ or $[t_j, r_j[$. The probability

$$\mathbb{P}\{X_u \in F \text{ for all } u \in \Delta(F) \cap I,\ X_t \notin F$$

is zero for all $t \in I$: if t is an inner point of I in the ordinary sense, use (27.2) ; if t is one of the left-hand end-points of intervals $[t_j, r_j[$, use (27.3).

To get the set D of the statement, possessing the above properties for all closed sets, it suffices to take the union of the countable sets $\Delta(F)$, F running through the countable set $\mathcal{H}$.

28 We come to Doob's main theorems, the first one concerning arbitrary processes and the second one measurable processes. We first give two examples.

- Let Ω consist of a single point. A process then is simply a function $f(t)$ on $\mathbb{R}_+$ which may be arbitrary bad. On the other hand (27.1) tells us that there exists some D such that

$$(f(t) \in F \text{ for } t \in D \cap I) \Leftrightarrow (f(t) \in F \text{ for } t \in I).$$

It follows that f is a right D-separable function and the "process" f therefore is right separable. So separability in itself doesn't imply any regularity of the sample functions of a process.

- We return to example (24.1). For every countable set D, we have $\mathbb{P}\{X_u = 0, u \in D\} = 1$, whereas for almost all ω the set $\{u : X_u(\omega) = 1\}$ is non-empty. Hence the process is not separable, and any attempt to make it separable would also make it indistinguishable from 0, and hence without interest.

29 THEOREM. <u>Every real-valued process</u> $(X_t)_{t \in \mathbb{R}_+}$ <u>has a right separable modification with values in</u> $\bar{\mathbb{R}}$ (1).

<u>Proof</u> : We fix $t \in \mathbb{R}_+$. Choosing the set D as in no. 27, we denote by $A_t(\omega)$ the - non-empty - set of cluster values in $\bar{\mathbb{R}}$ of the function $X_.(\omega)$ at the point t from the right and along D

$$A_t(\omega) = \bigcap_n \overline{\{X_u(\omega),\ u \in D \cap [t, t + 1/n[\}}.$$

The set of ω such that $X_t(\omega) \in A_t(\omega)$ is measurable. Let indeed d be a metric defining the topology of $\bar{\mathbb{R}}$; the condition $X_t(\omega) \in A_t(\omega)$ is equivalent to

$$\forall n > 0,\ \forall m > 0,\ \exists u \in [t, t + 1/n[\cap D,\ d(X_t(\omega), X_u(\omega)) < 1/m.$$

We now claim that $X_t(\omega) \in A_t(\omega)$ for almost all ω . Suppose indeed that $X_t(\omega) \notin A_t(\omega)$, and come back to the countable family $\mathcal{H}$ of closed sets of no. 27 ; there exists an element F of $\mathcal{H}$ containing $A_t(\omega)$ such that $X_t(\omega) \notin F$ and hence a number m such that $d(X_t(\omega), F) > 1/m$. If $F_m = \{x : d(x,F) \le 1/m\}$, we have for n sufficiently large $X_u(\omega) \in F_m$ for all $u \in D \cap [t, t + 1/n[$, because F_m is a <u>neighbourhood</u> of the set of cluster values at t along D. Consequently, for a suitable choice of n, m and $F \in \mathcal{H}$,

(1) But may be not in $\mathbb{R}$.

we have $\omega \in H(n,m,F)$, where this denotes the set

$$\{\omega : X_u(\omega) \in F_m \text{ for } u \in D \cap [t,t+1/n[,\ X_t(\omega) \notin F_m\} .$$

Since this event has probability zero by the choice of D, so does the union of the $H(n,m,F)$ (n, m integers, $F \in \mathcal{H}$) and we have seen that this union contains the set $\{X_t \notin A_t\}$.

To get the required modification, we finally set

$$X'_t(\omega) = X_t(\omega) \text{ if } X_t(\omega) \in A_t(\omega),\ X'_t(\omega) = \liminf_{s \downarrow t, s \in D} X_s(\omega) \text{ otherwise } ^{(1)}.$$

Doob's second theorem concerns the existence of modifications of a process which are both right separable and progressive.

THEOREM. Let $(X_t)_{t \in \mathbb{R}_+}$ be a process on $(\Omega,\mathcal{F},\mathbb{P})$ with values in $\overline{\mathbb{R}}$, and let $\dot{X}_t$ be the class of the random variable X_t considered as an element of L^0, the space of classes of real-valued random variables on Ω with the metric of convergence in probability. 30

Then (X_t) has a measurable modification, if and only if the mapping $t \mapsto \dot{X}_t$ is a uniform limit (in L^0) of measurable step functions.

If this condition is satisfied (X_t) has a right separable and measurable modification. More precisely, if it is satisfied and if (X_t) is adapted to a filtration $(\mathcal{F}_t)$, the modification can be chosen to be right separable and progressively measurable with respect to the family $(\mathcal{F}_{t+})$.

Proof : It can easily be shown that the condition of the statement is equivalent to the following : $t \mapsto \dot{X}_t$ is measurable in the usual sense (i.e. the inverse image of every Borel set of L^0 is Borel in $\mathbb{R}_+$) and takes its values in a separable subset of L^0. This condition is the correct definition of measurability to be used, for example in the theory of integration with values in Banach spaces.

The space $\overline{\mathbb{R}}$ is homeomorphic to the interval $I = [-1,+1]$. We use this homeomorphism to replace $\overline{\mathbb{R}}$ by I and convergence in probability by convergence in norm in L^1. Throughout the rest of this proof L^0 is thus replaced by L^1.

(a) We suppose that (X_t) is measurable and show that the above condition is satisfied. Let $\mathcal{H}$ be the set of real-valued measurable processes (Y_t) on $(\Omega,\mathcal{F},\mathbb{P})$ such that : (Y_t) is uniformly bounded and the mapping $t \mapsto \dot{Y}_t$ of $\mathbb{R}_+$ into L^1 is Borel with values in a separable subset of L^1. Clearly all processes of the form

$$Y_t(\omega) = \sum_{k \in \mathbb{N}} I_{[k2^{-n},(k+1)2^{-n}[}(t) Y^k(\omega)$$

where n runs through $\mathbb{N}$ and the Y^k are arbitrary uniformly bounded random variables, form an algebra contained in $\mathcal{H}$ which generates the σ-field $\mathcal{B}(\mathbb{R}_+) \times \mathcal{F}$. On the other

(1) This is where the values $\pm\infty$ may occur.

hand, $\mathcal{H}$ is closed under monotone bounded convergence. The monotone class theorem then implies that every bounded measurable process (X_t) belongs to $\mathcal{H}$, and the condition of the statement follows from 1.17.

(b) Conversely, let (X_t) be a process with values in I, satisfying the above condition and adapted to $(\mathcal{F}_t)$ (if no family is given, take $\mathcal{F}_t = \mathcal{F}$ for all t). We consider elementary processes (Z^n_t) such that $\|X_t - Z^n_t\|_1 \leq 2^{-n}$ for all t. We can write

$$Z^n_t(\omega) = \sum_k I_{A^n_k}(t) H^n_k(\omega) \tag{29.1}$$

where the A^n_k form a partition of $\mathbb{R}_+$ and the H^n_k are random variables with values in I. We begin by turning the (Z^n_t) into processes adapted to the family $(\mathcal{F}_{t+})$.

Let s^n_k be the infimum of A^n_k and let (t_i) be a decreasing sequence of elements of A^n_k converging to s^n_k. The sequence may be constant if $s^n_k \in A^n_k$. Since the random variables X_{t_i} are uniformly bounded, we can suppose - replacing (t_i) by a subsequence if necessary - that the X_{t_i} converge weakly in L^1 to a $\mathcal{F}_{s^n_k+}$-measurable random variable L^n_k(1) (cf. II.25). We have $\|X_{t_i} - H^n_k\|_1 \leq 2^{-n}$ and hence also $\|L^n_k - H^n_k\|_1 \leq 2^{-n}$, since the norm, being the upper envelope of a family of linear functionals, is a l.s.c. function under the weak topology of L^1. Then the process

$$Y^n_t(\omega) = \sum_k I_{A^n_k(t)} L^n_k(\omega) \tag{29.2}$$

is progressive with respect to the family $(\mathcal{F}_{t+})$, and $\|X_t - Y^n_t\|_1 \leq 2.2^{-n}$ for all n and t. We set

$$Y_t(\omega) = \liminf_n Y^n_t(\omega). \tag{29.3}$$

This process still is progressive. On the other hand, for each fixed t, Y^n_t converges a.s. (II.10) to (X_t) and hence (Y_t) is a modification of (X_t).

(c) This modification is not yet right separable. We return to the set D from no. 27 - relative to (X_t) or (Y_t), this amounts to the same, since they are modifications of each other - and set as in no. 29.

$$A^n_t(\omega) = \overline{\{Y_u(\omega),\ u \in D \cap [t,t+1/n[\}},\quad A_t(\omega) = \bigcap_n A^n_t(\omega).$$

Let d be the usual metric on I. The process $d(Y_t(\omega), A^n_t(\omega)) = \sup_{s\in D} d(Y_t(\omega), Y_s(\omega)) I_{]s-1/n,s]}(t)$ is progressive with respect to the family $(\mathcal{F}_{t+1/n})$; hence the process $d(Y_t, A_t)$ is progressive with respect to $(\mathcal{F}_{t+})$. It only remains to define as in no. 29

---

(1) We are indebted to Hoffmann-Jørgensen for the correction of an error at this point in the first edition.

$$X'_t = Y_t \quad \text{if } d(Y_t, A_t) = 0, \text{ i.e. if } Y_t \in A_t$$
$$= \liminf_{s \downarrow t\ s \in D} Y_t \quad \text{otherwise.}$$

This is the required modification.

REMARKS. (1) Let $\mathcal{G}$ be the σ-field generated by the X_t, $t \in \mathbb{R}_+$. If the mapping $t \mapsto \dot{X}_t$ with values in $L^0(\mathcal{F})$ is Borel and takes its values in a separable subset, it satisfies the same condition relative to $L^0(\mathcal{G})$ and the above proof shows that there exist step processes of type (29.1)

(29.4) $$Z^n_t = \sum_k I_{A^n_k}(t) H^n_k(\omega)$$

where the H^n_k are $\mathcal{G}$-measurable and converge uniformly to (X_t) in probability. By theorem I.18, each random variable H^n_k admits a representation

(29.5) $$H^n_k = h^n_k((X_{t^{nk}_p})_{p \in \mathbb{N}})$$

where $(t^{nk}_p)_{p \in \mathbb{N}}$ is a sequence in $\mathbb{R}_+$ and h^n_k is a Borel function on $\overline{\mathbb{R}}^{\mathbb{N}}$. But then it is easy to see that the property that the (Z^n_t) given by (29.4) and (29.5) converge uniformly in probability to (X_t) depends only on the time law of the process (X_t). In other words, the existence of a measurable modification is a property of a process which depends only on its time law (1).

(2) We keep the above notation. The process (X_t), if it is $(\mathcal{B}(\mathbb{R}_+) \times \mathcal{F})$-measurable, has a $(\mathcal{B}(\mathbb{R}_+) \times \mathcal{G})$-measurable modification (X'_t). But the σ-field $\mathcal{G}'$ generated by the X'_t may be strictly contained in $\mathcal{G}$ and one cannot be sure that (X'_t) is $(\mathcal{B}(\mathbb{R}_+) \times \mathcal{G}')$-measurable (that (X'_t) is "naturally measurable"). The same kind of difficulties arise with progressive processes.

Random closed sets, stopping times

So far we have studied the properties of a process $(X_t)_{t \in \mathbb{R}_+}$ which can be deduced from the knowledge of $(X_t)_{t \in D}$, where D is countable and dense. We now suppose that (X_t) is measurable (more precisely, progressively measurable with respect to a family $(\mathcal{F}_t)$, the case where no family is given corresponding to the deterministic filtration $\mathcal{F}_t = \mathcal{F}$) and study directly the behaviour of paths on $\mathbb{R}_+$, using the theory of capacities - in this case theorem III.44.

DEFINITION. Let $(\Omega, \mathcal{F}, \mathbb{P})$ be a probability space with a filtration $(\mathcal{F}_t)_{t \in \mathbb{R}_+}$ and let A be a subset of $\mathbb{R}_+ \times \Omega$. We denote by (a_t) the indicator process of A 31

(31.1) $$a_t(\omega) = I_A(t, \omega).$$

---

(1) Another such property is given in n° 46. About the existence of measurable modifications, see Hoffmann-Jørgensen, Z. Warsch. 25, 1973, p. 205-207.

A *is called* a measurable random set *if* $A \in \mathcal{B}(\mathbb{R}_+) \times \mathcal{F}$ (i.e. *if the process* (a_t) *is measurable*), *a* progressive set *if* (a_t) *is a progressive process, a* closed (*resp.* right, left, closed) *set if for all* ω *the section* $A(\omega) = \{t \in \mathbb{R}_+ : (t,\omega) \in A\}$ *is closed* (*resp. right, left closed*) *in* $\mathbb{R}_+$.

It may be useful to recall here that a "right-closed interval" $]a,b]$ or $[a,b]$ of $\mathbb{R}_+$ is closed under the *left* topology of $\mathbb{R}_+$ and is hence a *left-closed* set.

The progressive sets form a σ-field on $\mathbb{R}_+ \times \Omega$, the *progressive* σ-*field* ; the progressive processes are precisely the functions on $\mathbb{R}_+ \times \Omega$ which are measurable with respect to this σ-field.

Given a subset A of $\mathbb{R}_+ \times \Omega$, the closure $\bar{A}$ of A is the set whose section $\bar{A}(\omega)$ is, for all $\omega \in \Omega$, the closure of the section $A(\omega)$. We define similarly the right or left closure of A. The following proposition is the key to the results of this section.

32 THEOREM. *Suppose that the space* $(\Omega,\mathcal{F},\mathbb{P})$ *is complete, that* $\mathcal{F}_0$ *contains all the negligible sets and that the family* $(\mathcal{F}_t)$ *is right-continuous. Then the closure* (*resp. right, left closure*) *of a progressive set is a progressive set.*

Proof : First let A be a measurable random set. For all $s \geq 0$, we define

(32.1) $$D_s(\omega) = \inf\{t > s : (t,\omega) \in A\} \quad (\inf \emptyset = +\infty).$$

Clearly $D_{\cdot}(\omega)$ is a right-continuous increasing function on $\mathbb{R}_+$. On the other hand, by III.44, D_s is measurable with respect to the completion of $\mathcal{F}$ - that is, with respect to $\mathcal{F}$ itself. By 15, the process (D_s) is measurable and similarly so is the process $(D_{s-})_{s>0}$.

The set of cluster points of A under the right topology then is $\{(s,\omega) : D_s(\omega) = s\}$ and this set is measurable ; so is then the right closure of A. Similarly, the left closure of A is the set $\{(s,\omega) : s > 0, D_{s-}(\omega) = s\} \cup A$ (1), which is measurable. Taking their union, the closure $\bar{A}$ of A is measurable.

Progressivity follows immediately : what we have done on $[0,\infty[\times \Omega$ carries over to $[0,t[\times \Omega$ using an increasing bijection of $[0,\infty[$ onto $[0,t[$, and it follows that the closure of $A \cap ([0,t[\times \Omega)$ in $[0,t[\times \Omega$ is measurable relative to the σ-field $\mathcal{B}([0,t[) \times \mathcal{F}_t$. But then the remarks of no. 14 imply that A is progressive with respect to the family $\mathcal{F}_{t+} = \mathcal{F}_t$.

Here is a stronger version of 32 :

---

(1) The addition of A is necessary only to take account of the instant $t = 0$.

THEOREM. *With the same hypotheses on the space and family of σ-fields as in no.* 32, 33
let (X_t) *be a real-valued progressive process. Then the following processes are progressive* :

(a) $X_t^* = \sup_{s \le t} X_t$

(b) $\bar{Y}_t^+ = \limsup_{s \downarrow t} X_s$, $\underline{Y}_t^+ = \liminf_{s \downarrow t} X_s$ *and the analogous processes on the left-hand side.*

(c) $\bar{Z}_t^+ = \limsup_{s \downarrow\downarrow t} X_s$, $\underline{Z}_t^+ = \liminf_{s \downarrow\downarrow t} X_s$ *and the analogous processes on the left-hand side.*

Proof : (a) We write $L_0 = -\infty$, $L_t = \sup_{s<t} X_s$ for $t > 0$. As $X_t^* = L_t \vee X_t$, it suffices to show that (L_t) is progressive. But this process is left-continuous : by 15, it suffices to show that L_t is $\mathcal{F}_t$-measurable. Now the set $\{L_t > a\}$ is the projection on Ω of the $(\mathcal{B}(\mathbb{R}_+) \times \mathcal{F}_t)$-measurable set

$$\{(s,\omega) : s < t,\ X_s(\omega) > a\}.$$

So it is $\mathcal{F}_t$-analytic and hence $\mathcal{F}_t$-measurable, $\mathcal{F}_t$ being complete

(b) and (c) we deal only with $\bar{Y}^+$ and $\bar{Z}^+$. As $\bar{Y}_t^+ = X_t \vee \bar{Z}_t^+$, it suffices to consider only the process $\bar{Z}^+$. But we note that $\bar{Z}_t^+(\omega) \ge a$ if and only if, for all $\varepsilon > 0$, t is a right cluster point of $\{s : X_s(\omega) > a - \varepsilon\}$. Then we denote by A the set $\{(s,\omega) : X_s(\omega) > a - \varepsilon\}$ and return to the discussion of no. 32 : we have seen that the set of right cluster points of A can be written as $\{(s,\omega) : D_s(\omega) = s\}$ and that it is measurable (and progressive by the end of the proof of 32). It follows that $(\bar{Z}_t^+)$ itself is progressive.

> One can say more about the processes constructed above from left-hand limits, but we must wait until the *predictable* σ-field has been defined.

REMARK. Contrary to what happens in the theory of separability, we have no general method to compute probabilities relative to the processes defined in no. 33.

Compare the following result to Theorem 18.

THEOREM. *Let* $(\Omega,\mathcal{F})$ *be a measurable space and let* (X_t) *be a measurable process with* 34
values in a separable metrizable space E. *Let* Ω_{rl}, Ω_r, Ω_c *be the subsets of* Ω *consisting of the* ω *whose paths are r.c.l.l., resp. right continuous, continuous.*

If E *is cosouslin, the complements of these three sets are* $\mathcal{F}$-*analytic* (and hence belong to the universal completion σ-field of $\mathcal{F}$).

Proof : We deal for example with Ω_r. We choose a countable dense subset D of $\mathbb{R}_+$ and return to the proof of 18 : E is imbedded in the cube $I = [0,1]^{\mathbb{N}}$, to which we adjoin an isolated point α. As in no. 18, we define

$$X_{t+}(\omega) = \lim_{s \downarrow t,\, s \in D} X_s(\omega) \text{ if this limit exists in } I$$
$$= \alpha \text{ otherwise.}$$

This process is measurable. To say that the path $X_{\cdot}(\omega)$ is right continuous on $\mathbb{R}_+$ amounts to saying that $X_t(\omega) = X_{t+}(\omega)$ for all t. Thus, Ω_r^c is the projection of the $(\mathcal{B}(\mathbb{R}_+) \times \mathcal{F})$-measurable set $\{(t,\omega) : X_t(\omega) \neq X_{t+}(\omega)\}$, hence it is $\mathcal{F}$-analytic.

Almost-equivalence, almost-modifications

Nos. 35 to 45 can be omitted at a first reading ; nos. 39-45 will not be used later in this book.

35 We consider on a probability space $(\Omega,\mathcal{F},\mathbb{P})$ a process $(X_t)_{t \in \mathbb{R}_+}$ taking its values in a metrizable separable space E. We assume that the mapping $(t,\omega) \mapsto X_t(\omega)$ of $\mathbb{R}_+ \times \Omega$ into E is measurable, where $\mathbb{R}_+ \times \Omega$ is given the completed σ-field of $\mathcal{B}(\mathbb{R}_+)\times\mathcal{F}$ with respect to the measure $dt \otimes dP(\omega)$, a property which we express by saying (X_t) is Lebesgue measurable. Throughout this section we adopt the point of view of no. 5, according to which we do not have access to the r.v. X_t themselves, but only to functions on Ω of the form

(35.1) $$M_\phi^X(\omega,g) = \int_0^\infty g(X_t(\omega))\phi(t)dt$$

where g is Borel on E and positive or bounded and ϕ is Borel, positive and integrable on $\mathbb{R}_+$. Measurability in the Lebesgue sense guarantees both the existence of integrals (35.1) and the fact that the functions $M_\phi^X(.,g)$ are r.v. on the completed space $(\Omega,\mathcal{F})$.

DEFINITION. Two processes $(X_t)_{t \in \mathbb{R}_+}$, $(Y_t)_{t \in \mathbb{R}_+}$ with values in E (defined on possibly different probability spaces and Lebesgue measurable, are said to be almost-equivalent if, for every finite system of pairs (ϕ_i,g_i) $(1 \leq i \leq n$; ϕ_i Borel, positive and integrable on $\mathbb{R}_+$ and g_i Borel and positive on E), the following random variables with values in $\mathbb{R}^n$ have the same law

(35.2) $$(M_{\phi_1}^X(.,g_1),\ldots, M_{\phi_n}^X(.,g_n)),\ (M_{\phi_1}^Y(.,g_1),\ldots, M_{\phi_n}^Y(.,g_n)).$$

This is true in particular if (X_t) and (Y_t) are defined on the same probability space and if

(35.3) for almost all t $X_t = Y_t$ a.s.

We then say that these processes are almost-modifications of each other.

Our aim now consists in describing some properties of processes which depend only on their almost-equivalence class, and choosing best possible almost-modifications.

REMARK. To define the r.v. (35.1), it is not necessary that X_t be defined for all t ; everything extends to processes $(X_t)_{t \in T}$, where T is a set of full Lebesgue measure in $\mathbb{R}_+$. We leave aside such trivial extensions.

The essential topology on $\mathbb{R}$.

It is quite natural, in studying almost-equivalences, to try to use topologies on $\mathbb{R}$ which ignore sets of measure zero. This has already been done in analysis at the beginning of the 20 th Century. Many results in the theory of differentiation depend on the notion of "approximate limit", or limit along a set of density 1, a procedure which was also introduced into probability theory by Ito [1]. But we owe to Doob, Chung and Walsh the systematic study of these rather strange topologies on the line, and the discovery that another topology, the essential topology, is better adapted to the needs of probability theory. We define it by the corresponding limit notion, which is not quite correct, but allows a much more intuitive description.

We recall that if f is a real-valued function on $\mathbb{R}$ and I is an interval, $\operatorname{ess\,sup}_{s\in I} f(s)$ denotes the greatest lower bound of the numbers c such that the set $\{s : s \in I, f(s) > c\}$ is negligible relative to Lebesgue measure. This definition does not require f to be measurable. We do not change $\operatorname{ess\,sup}_{s\in I} f(s)$ by modifying f on a set of measure zero. As $\operatorname{ess\,sup}_{s\in I} f(s)$ is an increasing function of I, we can make the following definitions.

36 DEFINITION. Let f be a real-valued function on $\mathbb{R}$, measurable or not, and let $f \in \mathbb{R}$. We define [1]

(36.1) $$\operatorname{ess\,lim\,sup}_{s\downarrow\downarrow t} f(s) = \lim_{\varepsilon\downarrow\downarrow 0} \operatorname{ess\,sup}_{t<s<t+\varepsilon} f(s)$$

(36.2) $$\operatorname{ess\,lim\,sup}_{s\to t, s\neq t} f(s) = \lim_{\varepsilon\downarrow\downarrow 0} \operatorname{ess\,sup}_{t-\varepsilon<s<t+\varepsilon} f(s)$$

(36.3) $$\operatorname{ess\,lim\,sup}_{s\to t} f(s) = f(t) \vee \operatorname{ess\,lim\,sup}_{s\to t, s\neq t} f(s).$$

We leave to the reader the task of defining $\operatorname{ess\,lim\,sup}_{\ldots}$, where... represents "$s \uparrow\uparrow t$", "$s \downarrow t$", "$s \uparrow t$", and also of defining the corresponding lim inf, and the true essential limits if they exist.

The limit (36.2), for example, can be written $\limsup_{s\to t, s\neq t} f(s)$ relative to a true topology on $\mathbb{R}$, the essential topology, which can be described as follows ; an essential neighbourhood of t is a (not necessarily measurable) set containing $\{t\}$ and a set $]t-\varepsilon, t+\varepsilon[\setminus N$, where ε is > 0 and N is a negligible Borel set. Similarly, (36.1) can be written $\limsup_{s\to t, s\neq t} f(s)$ under another topology on R, the right essential topology, whose twin sister is the left essential topology. Here we leave details to the reader.

We group the fundamental properties of essential limits into two statements, the first of which concerns functions on $\mathbb{R}$ and the second processes. From the logical

(1) See footnote to no. 17

point of view, the first one is a special case of the second, but we give a more detailed proof of the first statement.

37 THEOREM. For every function f on $\mathbb{R}$ we denote by $\bar{f}$ the function $t \mapsto \text{ess} \limsup_{s \downarrow\downarrow t} f(s)$.

(a) If $f = g$ a.e., then $\bar{f} = \bar{g}$ everywhere.

(b) The function $\bar{f}$ is Borel and

$$\bar{f}(t) = \text{ess} \limsup_{s \downarrow\downarrow t} \bar{f}(s) = \limsup_{s \downarrow\downarrow t} \bar{f}(s). \tag{37.1}$$

(c) $f \leq \bar{f}$ almost everywhere on $\mathbb{R}_+$.

(d) If f is continuous under the right essential topology, f is right continuous under the ordinary topology.

Proof : We begin by establishing a lemma (which amounts to saying that the right topology on $\mathbb{R}_+$ has the (LL) property of no. III. 63).

Let $(L_i)_{i \in I}$ be a family of intervals $L_i = [a_i, b_i[$. There exists a countable subfamily $(L_i)_{i \in J}$ such that $\bigcup_{i \in I} L_i = \bigcup_{i \in J} L_i$.

We write $L = \bigcup_{i \in I} L_i$, $K_i =]a_i, b_i[$, $K = \bigcup_{i \in I} K_i$. Since the usual topology on the line has a countable base, there exists a countable subset J_1 of I such that $\bigcup_{i \in J_1} K_i = K$. On the other hand, the set $L \setminus K$ is countable, since it is discrete under the right topology : if a belongs to $L \setminus K$, there exists an interval $I = [a,b[\in (L_i)$, then every point of $]a,b[$ belongs to K and hence $I \cap (L \setminus K) = \{a\}$. Let then J_2 be a countable subset of I such that $L \setminus K \subset \bigcup_{i \in J_2} L_i$. It only remains to set $J = J_1 \cup J_2$.

We deduce, for the right essential topology, the following consequence : if $I \subset \mathbb{R}_+$ is discrete under the right essential topology, it has outer Lebesgue measure zero. We may indeed associate with every $t \in I$ an interval $L_t = [t, t+\varepsilon[$ such that $L_t \cap I$ has outer measure zero. According to the lemma, we can cover I with a countable infinity of such intervals, and this implies that I has outer measure zero.

We now prove theorem 37. Property (a) is obvious. Similarly, (c) is obvious since the set $\{f \geq \bar{f} + \varepsilon\}$ is discrete under the right essential topology for all $\varepsilon > 0$. To show that $\bar{f}$ is Borel, we note that the property $(\bar{f}(t) \geq a)$ means that, for all $\varepsilon > 0$ and all $r > 0$, the set $[t, t+r[\cap \{f > a - \varepsilon\}$ has strictly positive outer measure. If follows immediately that the set $\{\bar{f} \geq a\}$ is closed under the right topology and hence Borel, and that $\bar{f}$ is Borel. One also sees that $\bar{f}$ is u.s.c. under the right topology and hence that

$$\bar{f}(t) \geq \limsup_{s \downarrow t} \bar{f}(s) \geq \limsup_{s \downarrow\downarrow t} \bar{f}(s) \geq \text{ess} \limsup_{s \downarrow\downarrow t} \bar{f}(s)$$

$$\geq \text{ess} \limsup_{s \downarrow\downarrow t} f(s) = \bar{f}(t)$$

where the second row of inequalities arises from the fact that $f \le \bar{f}$ a.e. This establishes (37.1).

If f is continuous under the right essential topology, then $f = \bar{f}$ and hence f is u.s.c. under the right topology. Applying this result to -f, we see that f is right continuous.

REMARK. There is another way of showing that $\bar{f}$ is Borel, when f is Lebesgue measurable. It is well known that, on every probability space and for every random variable X,

$$\| X \|_\infty = \lim_{p\to\infty} \| X \|_p = \lim_p \mathbb{E}[|X|^p]^{1/p}.$$

Here this gives, for positive and bounded f,

(37.2) $$\operatorname*{ess\,sup}_{t<s<t+\varepsilon} f(s) = \lim_n \left(\frac{1}{\varepsilon}\int_t^{t+\varepsilon} f^n(s)ds\right)^{1/s}.$$

The function of t on the right-hand side is continuous and increases with n. The left-hand side is therefore a l.s.c. function under the ordinary topology and $\bar{f}$ is a Baire function of the second class.

THEOREM. Let (X_t) be a real-valued process on $(\Omega,\mathcal{F},\mathbb{P})$, adapted to the family $(\mathcal{F}_t)$ and Lebesgue measurable. We set 38

(38.1) $$\bar{X}_t^+(\omega) = \operatorname*{ess\,lim\,sup}_{s\downarrow\downarrow t} X_s(\omega),\quad \underline{X}_t^+(\omega) = \operatorname*{ess\,lim\,inf}_{s\downarrow\downarrow t} X_s(\omega).$$

(a) The process $(\bar{X}_t)$ is indistinguishable from a measurable process (in the ordinary sense) and is even progressive with respect to $(\mathcal{F}_{t+})$.

(b) (X_t) has a right-continuous almost-modification if and only if $(\bar{X}_t^+)$ and $(\underline{X}_t^+)$ are indistinguishable.

(c) If $(\Omega,\mathcal{F},\mathbb{P})$ is complete, the set of ω such that $X_.(\omega)$ is equal a.e. to a right continuous function is measurable.

Proof : (a) Since the process (X_t) is Lebesgue measurable, there exist two processes (U_t), (V_t), measurable in the ordinary sense, such that $U_t \le X_t \le V_t$ and that $\{(t,\omega) : U_t(\omega) < V_t(\omega)\}$ is negligible relative to the measure $dt \otimes d\mathbb{P}(\omega)$. By Fubini's Theorem, the set A of ω such that $U_.(\omega) = V_.(\omega)$ does not hold a.e. is $\mathbb{P}$-negligible and, if $\omega \notin A$, then $\bar{U}_t^+(\omega) = \bar{X}_t^+(\omega) = \bar{V}_t^+(\omega)$ for all t by 37 (a). Hence it suffices to show that the process $(\bar{U}_t^+)$, for example, is progressive with respect to the family $(\mathcal{F}_{t+})$. To this end, we reduce the problem to the case where (U_t) takes its values in the interval [0,1] and note that the process

$$t \mapsto \left(\frac{1}{\varepsilon}\int_t^{t+\varepsilon} U_s^n ds\right)^{1/n}$$

has continuous paths and is adapted to the family $(\mathcal{F}_{t+\varepsilon})$: hence it is progressive with respect to this family. As $n \to \infty$, we deduce that the process $t \mapsto \operatorname*{ess\,sup}_{t<s<t+\varepsilon} U_s$ is

progressive with respect to the same family and, as $\varepsilon \to 0$, it follows that $(\bar{U}_t^+)$ is progressive with respect to the family $(\mathcal{F}_{t+})$ (no. 14).

(b) Suppose that (X_t) has a right continuous almost-modification (Y_t). By Fubini's Theorem, there exists a measurable set $N \subset \Omega$ such that $\mathbb{P}(N) = 0$ and, for $\omega \notin N$, $X_.(\omega) = Y_.(\omega)$ a.e. Then, for $\omega \notin N$, $\bar{X}_t^+(\omega) = \bar{Y}_t^+(\omega) = Y_t(\omega) = \underline{Y}_t^+(\omega) = \underline{X}_t^+(\omega)$ and the two processes are indistinguishable. Conversely, if $\bar{X}_.^+(\omega) = \underline{X}_.^+(\omega)$, their common value is a function which is both u.s.c. and l.s.c. under the right essential topology, hence essentially right continuous, therefore right continuous (37,(d)) and a.e. equal to $X_.(\omega)$ (37(c)). If the processes $(\bar{X}_t^+)$ and $(\underline{X}_t^+)$ are indistinguishable, there exists a measurable set N such that $\mathbb{P}(N) = 0$, and for $\omega \notin N$, $\underline{X}_.^+(\omega) = \bar{X}_.^+(\omega)$; the required almost-modification (Y_t) can be taken to be the common value of these processes on N^c and to 0 on N. It must be noted that it is not <u>progressively</u> measurable ; it becomes so if the set N of measure zero is adjoined to $\mathcal{F}_0$.

(c) We leave to the reader the task of verifying, as for (b), that the set in the statement is equal to $\{\omega : \bar{X}_.^+(\omega) = \underline{X}_.^+(\omega)\}$. Its complement is the projection of $\{(t,\omega) : \bar{X}_t^+(\omega) \neq \underline{X}_t^+(\omega)\}$; since this set is equal to an element of $\mathcal{B}(\mathbb{R}_+) \times \mathcal{F}$ up to an evanescent set (a), its projection is $\mathcal{F}$-measurable (III. 13 and III. 33).

39 REMARK. This projection argument gives the following more precise result : we make no completion hypothesis and suppose that (X_t) is measurable. Then the set of $\omega \in \Omega$ such that $X_.(\omega)$ is equal a.e. to a right continuous function is the complement of an $\mathcal{F}$-analytic set. It is natural to ask, by analogy with no. 18, <u>whether the set of ω such that $X_.(\omega)$ is equal a.e. to a r.c.l.l. mapping is $\mathcal{F}$-measurable</u>. It is indeed so and we sketch a proof. First an interesting notion :

DEFINITION. <u>Let</u> A <u>be a measurable set of</u> $\mathbb{R}_+ \times \Omega$. <u>The</u> essential debut <u>of</u> A <u>is the function on</u> Ω

$$E_A(\omega) = \inf\{t : \int_0^t I_A(s,\omega)ds > 0\} \qquad (\inf \emptyset = +\infty).$$

The essential debut thus appears as the ordinary debut of the measurable set $\{(t,\omega) : \int_0^t I_A(s,\omega) > 0\}$.

LEMMA. <u>The essential debut of a measurable set</u> A <u>is</u> $\mathcal{F}$-<u>measurable</u>.

<u>Proof</u> : This follows immediately from the equality, true for all $t \geq 0$,

$$\{\omega : E_A(\omega)\} \geq t\} = \{\omega : \int_0^t I_A(s,\omega)ds = 0\}.$$

The proof of the property stated above is now quite similar to that of no. 18 ; simply, the times T_n^ε must be replaced by essential debuts.

Pseudo-paths

40 Let $(\Omega,\mathcal{F},\mathbb{P})$ be a probability space and E a separable metrizable state space - which we shall suppose to be <u>Lusin</u> throughout this section, in order to avoid uninteresting

complications. We can then identify E, as a measurable (but not as a topological) space, to a Borel subset of the interval $I = [0,1]$: this permits us to use the tools of real analysis. We assume that $0 \in E$, but the role of 0 below could be played by any point of E arbitrarily chosen.

We are going to study classes of Lebesgue measurable processes $(X_t)_{t \in \mathbb{R}_+}$ with values in E under the almost-equivalence relation defined in no. 35, and in particular to construct a canonical representative of each almost-equivalence class.

A first remark : we may consider a process (X_t) with values in E, defined on Ω and Lebesgue measurable, as a process with values in I. We know, by completion theory, that there exists a process (Y_t), measurable in the ordinary sense, such that the set $\{(t,\omega) : X_t(\omega) \neq Y_t(\omega)\}$ is negligible relative to the measure $dt \otimes d\mathbb{P}(\omega)$. Replacing Y_t by 0 if $Y_t \notin E$, we can suppose that (Y_t) takes its values in E. But then (Y_t) is an almost-modification of (X_t), by Fubini's Theorem. Hence there is no loss of generality, from the point of view of "almost-equivalence" classes, in confining attention to measurable processes in the ordinary sense.

DEFINITION. Let f be a Borel mapping of $\mathbb{R}_+$ into E. The image of the Lebesgue measure under the mapping $t \mapsto (t,f(t))$ from $\mathbb{R}_+$ to $\mathbb{R}_+ \times E$ is a measure, called the pseudo-path associated with f, and denoted by $\psi(f)$. 41

Let $(X_t)_{t \in \mathbb{R}_+}$ be a measurable process defined on Ω with values in E. The pseudo-path associated with the Borel mapping $X_{\cdot}(\omega)$ is called the pseudo-path of ω and is denoted by $\psi^X(\omega)$ or $\psi(\omega)$.

Our aim now is to provide the set of all pseudo-paths with a Lusin measurable structure such that the mapping ψ is measurable.

DEFINTION. We denote by $\mathcal{L}(E)$ the set of positive measures on $\mathbb{R}_+ \times E$, whose projection onto $\mathbb{R}_+$ is the Lebesgue measure, and by $\Pi(E) \subset \mathcal{L}(E)$ the set of pseudo-paths associated with all Borel mappings of $\mathbb{R}_+$ into E. These spaces are given the coarsest topology such that all mappings $\lambda \mapsto \langle \lambda, f \otimes g \rangle$ are continuous, where f is continuous with compact support on $\mathbb{R}_+$ and g bounded and continuous on E, and the corresponding measurable structure. 42

It is very easy to describe directly the measurable structure : by the monotone class theorem, it is also generated by the mappings $\lambda \mapsto \langle \lambda, j \rangle$, where j is positive and Borel on $\mathbb{R}_+ \times E$. Hence it depends only on the measurable structure on E and not on the topology chosen on E. We shall leave aside the topological structure (which might be studied by imbedding E as a topological subspace of a compact metric space).

THEOREM. If E is Lusin, $\Pi(E)$ is a Lusin measurable space. 43

Proof : We imbed E in I. $\mathcal{L}(I)$ is a set of Randon measures on the LCC space $\mathbb{R}_+ \times I$

and the topology of $\mathcal{L}(I)$ is the topology of vague convergence [1]. On the other hand, every measure $\mu \in \mathcal{L}(I)$ gives to $[0,n] \times I$ the measure n and hence $\mathcal{L}(I)$ is bounded under the vague topology. Since it is obviously closed, it is compact and metrizable. The function $\mu \mapsto \mu(E^c)$ is Borel on $\mathcal{L}(I)$ and hence the set $\mathcal{L}(E) = \{\mu \in \mathcal{L}(I) : \mu(E^c) = 0$ is Borel in $\mathcal{L}(I)$. On the other hand, $\Pi(E) = \Pi(I) \cap \mathcal{L}(E)$, for if f denotes a Borel mapping of $\mathbb{R}_+$ into I and if the image of the Lebesgue measure λ on $\mathbb{R}_+$ under $t \mapsto (t,f(t))$ is carried by $\mathbb{R}_+ \times E$, then $f(t) \in E$ for almost all t and there exists a Borel mapping of $\mathbb{R}_+$ into E giving the same image measure. Hence it suffices to show that $\Pi(I)$ is Borel in $\mathcal{L}(I)$.

To this end we show, following an idea of Mokobodzki, that $\Pi(I)$ is the set of extreme points of the metrizable compact convex set $\mathcal{L}(I)$. As we shall see in Chapter XI [2] that the set of extreme points of such a convex set is a $\mathcal{G}_\delta$, this will establish the required result.

Then let μ be the image of λ under $t \mapsto (t,f(t))$, where f is Borel with values in I and suppose that $\mu = c\mu_1 + (1-c)\mu_2$, with $c \in]0,1[$ and $\mu_1, \mu_2 \in \mathcal{L}(I)$. We disintegrate μ_1 and μ_2 with respect to their projection λ on $\mathbb{R}_+$

$$\mu_i(dt,dx) = \int \varepsilon_s(dt) \otimes \alpha_s^i(dx)\lambda(ds) \quad (i = 1,2)$$

where (α_s^i) is a measurable family of probability laws on I. Then for almost all s, $c\alpha_s^1 + (1-c)\alpha_s^2 = \varepsilon_{f(s)}$, hence $\alpha_s^1 = \alpha_s^2 = \varepsilon_{f(s)}$ and finally $\mu_1 = \mu_2 = \mu$: μ is indeed extremal.

Conversely, let μ be a measure which belongs to $\mathcal{L}(I)$ but not to $\Pi(I)$; we show that it is not extremal in $\mathcal{L}(I)$. We can write $\mu = \int \varepsilon_s \otimes \alpha_s\lambda(ds)$ as above. For all s, we denote by g(s) and h(s) the greatest lower bound and least upper bound of the support of α_s ; they are Borel functions of s (for example, g(s) is the supremum of the rationals r such that $\alpha_s([0,r[) > 0)$ and the fact that $\mu \notin \Pi(I)$ implies that $\{s : g(s) < h(s)\}$ is not negligible. We write $j = \frac{g+h}{2}$ and then, for $c \in]0,1[$,

$$m_s(dx) = \alpha_s(dx)I_{[0,j(s)[}(x) \,\|\alpha_s(I_{[0,j(s)[})\|$$
$$\text{if } \alpha_s(I_{[0,j(s)[}) > c$$
$$m_s(dx) = \alpha_s(dx) \quad \text{otherwise.}$$

The measure $\mu' = \int \varepsilon_s \otimes m_s\lambda(ds)$ belongs to $\mathcal{L}(I)$ and $cm_s \leq \alpha_s$, so that the measure $\mu'' = \mu - c\mu'/(1-c)$ is positive. Since the projection of μ'' on $\mathbb{R}_+$ is equal to the Lebesgue measure, we also have $\mu'' \in \mathcal{L}(I)$ and $\mu = c\mu' + (1-c)\mu''$. It only remains to note that if $g(s) < h(s)$, α_s is non-zero on both intervals $[0,j(s)[$ and $[j(s),1]$ and hence $m_s \neq \alpha_s$ if c is sufficiently small. For c = 1/n and n sufficiently large,

(1) The vague topology on an LCC space is studied in detail in Bourbaki [4], Integration, Chapter 3.

(2) Theorem XI. 24 of the first edition.

the representation $\alpha = c\mu' + (1-c)\mu''$ is therefore non-trivial and μ is not extremal.

REMARK.(a) The above argument can be applied to E imbedded in I, instead of I, and then shows that $\Pi(E)$ is always the set of extremal points of the (not necessarily compact) convex set $\mathcal{L}(E)$.

Suppose that E is Polish. We imbed it as a $\mathcal{G}_\delta$ in some compact metrizable space K : the topology of $\mathcal{L}(E)$ is induced by that of $\mathcal{L}(K)$ and $\mathcal{L}(E)$ is a $\mathcal{G}_\delta$ of $\mathcal{L}(K)$ (cf. III. 60). On the other hand, $\Pi(K)$ is also a $\mathcal{G}_\delta$ of $\mathcal{L}(K)$, by the above remark. Hence $\mathcal{L}(E) \cap \Pi(K) = \Pi(E)$ is also a $\mathcal{G}_\delta$ of $\Pi(K)$ and hence is a Polish space.

(b) There exists on $\Pi(E)$ a stronger topology than that of no. 42 inducing the same Borel structure. Let d be a bounded metric defining the topology of E and let μ and μ' be two elements of $\Pi(E)$ which are the images of the Lebesgue measure on $\mathbb{R}_+$ under two mapping $t \mapsto (t,f(t))$, $t \mapsto (t,f'(t))$. We then define on $\Pi(E)$ a metric $\bar{d}$ by

(43.1) $$\bar{d}(\mu,\mu') = \int_0^\infty d(f(t), f'(t))e^{-t}dt.$$

This amounts to considering $\Pi(E)$ as a set of classes of measurable mappings of $\mathbb{R}_+$ into E under the topology of convergence in measure. $\Pi(E)$ then is separable, since any element of $\Pi(E)$ can be approximated in measure, first by a continuous function, and then by a dyadic step function with values in a countable dense set. It is very easy to see that the topology of $\Pi(E)$ and the topology of convergence in measure lead to the same Borel σ-field. However, our first topology on $\Pi(E)$ may be better in some respects (it does not seem true that $\Pi(E)$ is Polish under the topology of convergence in measure if E is Polish).

We now consider on Ω a measurable process (X_t) with values in E. If ϕ is a continuous function with compact support on $\mathbb{R}_+$ and g is a bounded continuous function on E, clearly the function $\omega \mapsto \int g \circ X_s(\omega)\phi(s)ds = \langle\psi^X(\omega), \phi \otimes g\rangle$ is measurable on Ω . In other words, the mapping ψ of Ω into $\Pi(E)$ is measurable and we can consider the image law of $\mathbb{P}$ under ψ.

DEFINITION. Let (X_t) be a measurable process defined on $(\Omega,\mathcal{F},\mathbb{P})$, with values in a Lusin metrizable space E, the image law of P under the mapping ψ of Ω into $\Pi(E)$ is called the pseudo-law of (X_t). 44

REMARK. Let Ω be the set of right continuous (resp. r.c.l.l.) mappings of $\mathbb{R}_+$ into E, with the σ-field generated by the coordinate mappings. The mapping which associates with $\omega \in \Omega$ the corresponding pseudo-path $\psi(\omega)$ is injective : hence we can identify Ω with a subset of $\Pi(E)$. We have seen that ψ is measurable. On the other hand, if $\omega \in \Omega$ and f is bounded and continuous on E, then $f(\omega(t)) = \lim_{h\to 0}\int \frac{1}{h} f(\omega(t+s))ds$, so that the coordinate mappings on Ω are measurable relative to the σ-field induced by $\Pi(E)$; hence ψ is a Borel isomorphism and the results stated above (19(a)) according to which Ω is cosouslin (resp. Lusin if E is Polish) for its usual measurable structure, show that Ω is the complement of an analytic set (resp. a Borel set) in $\Pi(E)$.

We now show that the notion of a pseudo-law solves conveniently the almost-equivalence problem.

45 THEOREM. Let (X_t) and (Y_t) be two measurable processes with values in a Lusin space E (defined on possibly different probability spaces).

The following properties are equivalent

(a) (X_t) and (Y_t) have the same pseudo-law $\mathbb{Q}$;

(b) (X_t) and (Y_t) are almost-equivalent (no. 35) ;

(c) there exists a Borel subset $\mathbb{T}$ of $\mathbb{R}_+$ of full Lebesgue measure, such that the processes $(X_t)_{t \in \mathbb{T}}$ and $(Y_t)_{t \in \mathbb{T}}$ have the same time law.

Further, there exists a stochastic process (fixed once and for all) $(Z_t)_{t \in \mathbb{R}_+}$, defined on $\Pi(E)$ and measurable which has the following property :

For every law $\mathbb{J}$ on $\Pi(E)$, the pseudo-law of the process (Z_t) under $\mathbb{J}$ is $\mathbb{J}$ itself. In particular, if $\mathbb{J} = \mathbb{Q}$, (Z_t) belongs to the pseudo-equivalence class of (X_t).

Proof : We have already remarked that, if g is positive and Borel on E and ϕ is positive and Borel on $\mathbb{R}_+$,

$$\int g(X_t(\omega))\phi(t)dt = \langle\psi^X(\omega), \psi \otimes g\rangle$$

It follows immediately that (a) $\Leftrightarrow$ (b). We show that (a) $\Rightarrow$ (c). We again consider E as imbedded in $I = [0,1]$ with $0 \in E$. By the Lebesgue differentiation theorem, we have a.e. j denoting the identity mapping of E into $[0,1]$,

$$(45.1)\qquad X_t(\omega) = \lim_n n\int_t^{t+1/n} X_s(\omega)dt = \lim_n \langle\psi^X(\omega), e_n(t+.) \otimes j\rangle$$

where $e_n(s) = nI_{[0,1/n]}(s)$. We set, for every measure $\mu \in \Pi(E)$

$$(45.2)\qquad Z_t(\omega) = \liminf_n \langle\mu, e_n(t+.) \otimes j\rangle \text{ if this number belongs to E}$$

$$= 0 \quad \text{otherwise.}$$

Clearly the process (Z_t) on $\Pi(E)$ is measurable. On the other hand, by (45.1),

$$\text{for all } \omega,\ Z_.(\psi^X(\omega)) = X_.(\omega) \text{ a.e.}$$

and consequently, by Fubini's theorem, the set of all t such that $X_t = Z_t \circ \psi^X$ a.s. does not hold is negligible. Then let T denote the set of all t such that both $X_t = Z_t \circ \psi^X$ a.s. and $Y_t = Z_t \circ \psi^Y$ a.s. By the definition of image laws, the time law of the process $(X_t)_{t \in T}$ (resp. $(Y_t)_{t \in T}$) is the law of the process $(Z_t)_{t \in T}$ under $\mathbb{Q}$. Thus (a) implies (c).

We show that (c) implies (b). Replacing X_t and Y_t by 0 if $t \notin \mathbb{T}$, we may assume that both processes have the same time law on the whole of $\mathbb{R}_+$. Then it suffices to show the theorem on a finite interval $[0,A[$, which enables us to reduce it to the case where (X_t) and (Y_t) are periodic on $\mathbb{R}$ with period A. In the argument below we take $A = 1$.

We now follow Doob, Stochastic Processes, p. 63, to which we refer the reader for full detail. Let g be a bounded Borel function on E and ϕ be a bounded Borel function on $\mathbb{R}$ of period 1. We want to compute, for all ω belonging to the space Ω on which (X_t) is defined, the integral

$$M^X(\omega) = M^X_\phi(\omega,g) = \int_0^1 g(X_t(\omega)\phi(t)dt \tag{45.3}$$

and to this end we compare it to the following Riemann sum, where σ is a parameter belonging to $[0,1[$,

$$R^X_i(\sigma,\omega,\phi,g) = R^X_i(\sigma,\omega) = 2^{-i} \sum_{k<2^i} g(X_{\sigma+k2^{-i}}(\omega))\phi(\sigma + k2^{-i}).$$

It is known - an easy result from real analysis - that for all ω

$$\int |M^X(\omega) - R^X_i(\sigma,\omega)\, d\sigma \underset{i\to\infty}{\to} 0.$$

We integrate with respect to ω, thus getting convergence in L^1 relative to the measure $dP(\omega) \otimes d\sigma$, and choose a sequence (i_n) such that

$$R^X_{i_n}(\sigma,\omega) \to M^X(\omega) \text{ a.s. in } \sigma,\ \omega.$$

There then exists, by Fubini's theorem, a set N of measure zero such that for $\sigma \notin N$

$$R^X_{i_n}(\sigma,\omega) \to M^X(\omega) \text{ a.s. in } \omega\ . \tag{45.4}$$

Since these Riemann sums are uniformly bounded, convergence in fact takes place in every L^p $(p < \infty)$. Extracting a subsequence from (i_n) and enlarging the set N of measure zero if necessary, we can also assert that, on the space Ω' where (Y_t) is defined,

$$R^Y_{i_n}(\sigma,\omega') \to M^Y(\omega') \text{ a.s. in } \omega'. \tag{45.5}$$

This concerns a given pair (ϕ,g). But from successive extractions and enlargements of the set N we get (45.3) and (45.4) for any finite family (ϕ_i,g_i) of pairs as above and this permits us to compute the laws of the random vectors

$$(M^Y_{\phi_1}(.,g_1),\dots,\ M^Y_{\phi_n}(.,g_n)),(M^X_{\phi_1}(.,g_1),\dots,\ M^X_{\phi_n}(.,g_n))$$

using quantities which depend only on the time law, and which therefore are the same for both processes. Hence these are almost equivalent.

Finally, we leave to the reader the details concerning the last sentence of the statement : there is nothing more to prove, but it remains to state what was said about (Z_t) at the beginning of the proof in a slightly different way.

REMARK. We stated in no. 5 that the time law of a process is in general an imprecise notion, which gives access only to events depending on countably many instants. This is corrected by 45 in the case of measurable processes : two processes which are measurable and have the same law also have the same pseudo-law, and this allows a 46 lot of analysis on the sample functions. Let us prove a little more. Let $g_1,\dots,g_n$ be bounded Borel functions on the state space E of (X_t), $\mu_1,\dots,\mu_n$ be bounded measures on $\mathbb{R}_+$ (not necessarily absolutely continuous with respect to Lebesgue measure as in 45). We show that the time law competely determines the law of the random vector

$$U = \left(\int g_1 \circ X_s \, \mu_1(ds),\dots, \int g_n \circ X_s \, \mu_n(ds)\right).$$

Since U takes its values in a bounded set of $\mathbb{R}^n$, it suffices to check that we can compute $E[f(U)]$ for any polynomial $f : \mathbb{R}^n \to \mathbb{R}$. This reduces to the case of monomials, and finally to that of products

$$E\left[\left(\int h_1 \circ X_s \, \lambda_1(ds)\right)\dots\left(\int h_k \circ X_s \, \lambda_k(ds)\right)\right]$$

where k may be larger than n, due to the presence of powers of the coordinates in monomials. And now, thanks to Fubini's theorem, this is just

$$\int_{\mathbb{R}^k} \lambda_1(ds_1)\dots \lambda_k(ds_k) E[\, h_1 \circ X_{s_1} \dots h_k \circ X_{s_k}]$$

which depends only on the time law of (X_t). This proof, simpler than the original one, was suggested to us by Prof. T. Kurtz.

3. OPTIONAL AND PREDICTABLE TIMES

With this paragraph, we start developing the fundamental notions of the so called "general theory of processes" - at least those which can be studied without martingale theory. In spite of its name, this theory is not primarily concerned with processes, but rather with the structure determined on a measurable space $(\Omega,\mathcal{F})$ or a probability space $(\Omega,\mathcal{F},\mathbb{P})$, by a filtration $(\mathcal{F}_t)$.

The theory indeed has two slightly different forms. First the probabilistic theory, relative to a given probability measure $\mathbb{P}$, which is mainly developed under the hypotheses which we call below the usual conditions. On the other hand, the non-probabilistic theory (depending only on the space and not on any given law), whose importance has been shown by recent research, but which has not yet reached a definitive state. At a first reading, we advise the reader to omit all difficulties that arise when the usual conditions aren't assumed.

47 We refer the reader to nos. 11 to 15 for the first definitions about filtrations. We make here an additional convention : whenever we speak of a filtration $(\mathcal{F}_t)$; we assume that two additional σ-fields are given : one denoted by $\mathcal{F}_\infty$, containing all the $\mathcal{F}_t$, the other denoted by $\mathcal{F}_{0-}$ and contained in $\mathcal{F}_0$. We make the convention that $\mathcal{F}_t = \mathcal{F}_{0-}$ for $t < 0$ and $\mathcal{F}_{\infty-} = \bigvee_t \mathcal{F}_t$. This is very convenient and in no way restricts the generality : if such σ-fields aren't explicitly given, we may simply take $\mathcal{F}_{0-} = \mathcal{F}_0$, $\mathcal{F}_\infty = \mathcal{F}_{\infty-}$ (or $\mathcal{F}_\infty = \mathcal{F}$).

48 DEFINITION. *Let* $(\Omega,\mathcal{F},\mathbb{P})$ *be a probability space with a filtration* $(\mathcal{F}_t)_{t \in \mathbb{R}_+}$.

We say that the filtration is complete *if the probability space is complete* and if all the $\mathbb{P}$-negligible sets belong to $\mathcal{F}_{0-}$.

We say that the filtration satisfies the usual conditions if it is complete and right continuous (i.e. $\mathcal{F}_t = \mathcal{F}_{t+}$ for all $t \geq 0$).

An arbitrary filtration $(\mathcal{F}_t^o)$ can always be completed : one completes the space and then adjoins to each σ-field all the negligible sets. If this operation is performed on the family made right continuous $\mathcal{G}_t^o = \mathcal{F}_{t+}^o$ ($\mathcal{G}_{0-}^o = \mathcal{F}_{0-}^o$) one gets a family $(\mathcal{F}_t)$ which satisfies the usual conditions and which is called the usual augmentation of the family $(\mathcal{F}_t^o)$. We shall keep whenever possible the same notation principle as in the preceding sentence : the symbol ^{o} to denote any "raw" family of σ-fields (no usual conditions), then its suppression can be used to indicate that the usual augmentation has been performed.

Since all the negligible sets are adjoined to $\mathcal{F}_{0-}$, and not only the negligible sets of $\mathcal{F}_t^o$ to each $\mathcal{F}_t^o$, the operation of completion strongly affects the time structure of the family. For instance, in the theory of Markov processes, completion alone will turn the family $(\mathcal{F}_t^o)$ into a right continuous one (provided the semigroup is nice enough).

Optional times

Just as the seemingly trivial definition of the derivative contains in germ all of the Calculus, and its discovery may have involved as much genius as the whole developement that followed it, the seemingly trivial notion of stopping time (due to Doob) is the cornerstone of the "general theory of processes".

DEFINITION. Let $(\mathcal{F}_t^o)$ be a filtration on $(\Omega,\mathcal{F},\mathbb{P})$. A random variable T on Ω with values in $\mathbb{R}_+ \cup \{+\infty\}$ is called a stopping time or optional time (of the filtration $(\mathcal{F}_t^o)$) if 49

(49.1) for all $t \in \mathbb{R}_+$, the event $\{T \leq t\}$ belongs to $\mathcal{F}_t^o$.

Every positive constant is a stopping time.

We use both the older terminology (stopping time) and the more recent one (optional time or r.v.) with a slight preference to ward the latter (optional may be used as an adjective)!

REMARKS. (a) T is often called a wide sense stopping time if it satisfies the condition

(49.2) for all t, $\{T < t\} \in \mathcal{F}_t^o$.

But this is not really anything new: T is optional in the wide sense if and only if it is optional w.r. to the family $(\mathcal{F}_{t+}^o)$. If the family $(\mathcal{F}_t^o)$ is right-continuous, stopping times are characterized by (49.2), which is often much easier to verify than (49.1). Another characterization of wide sense stopping times is

(49.3) for all t, $T \wedge t$ is $\mathcal{F}_t^o$-measurable.

(b) In the case of a discrete filtration $(\mathcal{F}_n^o)_{n \in \mathbb{N}}$ stopping times are characterized by any one of the equivalent properties

(49.4) for all n, $\{T \le n\} \in \mathcal{F}_n^o$

(49.5) for all n, $\{T = n\} \in \mathcal{F}_n^o$.

Nothing analogous to (49.5) exists in general in the continuous case. But see in no. 94 the case of "canonical processes".

A fundamental example

Recall that the σ-field $\mathcal{F}_t^o$ is the set of events which we consider as prior to the instant t (no. 13), in other words known at time t in the universe of the observer. Let us imagine that the observer watches for the appearance of some random phenomenon and notes the (random) date $T(\omega)$ at which this phenomenon occurs for the first time. The event $\{T \le t\}$ means that the phenomenon has occurred at least once before t, hence it belongs to $(\mathcal{F}_t^o)$, and T is optional.

Here is the mathematical content of the preceding discourse : the proof tells a little more than the statement.

50 THEOREM. Assume that the family $(\mathcal{F}_t)$ satisfies the usual conditions and let A be a progressive subset of $\mathbb{R}_+ \times \Omega$ (31). Then the debut D_A of A (III. 44) is a stopping time.

Proof : We begin with the general case of a family $(\mathcal{F}_t^o)$ which does not satisfy the usual conditions. The set $\{D_A < t\}$ is the projection onto Ω of the set $\{(s,\omega) : s < t, (s,\omega) \in A\}$, which belongs to $\mathcal{B}(\mathbb{R}_+) \times \mathcal{F}_t^o$ by the definition of progressive sets (14). Then (cf. III. 44)

(50.1) for all t, $\{D_A < t\}$ is $\mathcal{F}_t^o$-analytic.

Under the hypotheses of the statement, the σ-field $\mathcal{F}_t$ is complete and hence equal to $\mathcal{A}(\mathcal{F}_t)$ (III. 33) and (49.1) implies that D_A is a stopping time of the family $(\mathcal{F}_{t+})$, which is equal to $(\mathcal{F}_t)$ by right-continuity.

51 REMARKS. (a) Under the usual conditions, every r.v. which is equal $\mathbb{P}$-a.s. to a stopping time is a stopping time and the theorem applies to every set A which is indistinguishable from a progressive set.

(b) In an arbitrary filtration $(\mathcal{F}_t^o)$, every stopping time T can be interpreted as the debut of the set $\{(t,\omega) : t > T(\omega)\}$, whose indicator is a right-continuous adapted, hence progressive, process (15). Similarly, every wide sense stopping time T is the debut of the progressive set $\{(t,\omega) : t > T(\omega)\}$ (on these sets, see 60).

(c) We return to the proof and suppose that A is right-closed. Then $\{D_A \le t\}$ is the projection of $\{(s,\omega) : s \le t, (s,\omega)\} \in A\}$ and, as for (50.1),

(50.2) for all t, $\{D_A \le t\}$ is $\mathcal{F}_t^o$-analytic.

(d) To illustrate the need for analytic sets in probability theory, let us give an example on measurability of debuts. Consider the set Ω of all r.c.l.l. mappings from $\mathbb{R}_+$ to $\mathbb{R}_+$. For every $\omega \in \Omega$ set $X_t(\omega) = \omega(t)$ and denote by $\mathcal{F}^o$ the σ-field generated by $(X_t)_{t \geq 0}$. Then $(\Omega, \mathcal{F}^o)$ is a nice Lusin measurable space, and the simplest hitting time of a closed set in $\mathbb{R}_+$

$$D(\omega) = \inf\{t : X_t(\omega) = 0\}$$

is such that the sets $\{D \leq T\}$ are analytic and not Borel. Of course the hitting times $D_\varepsilon(\omega) = \inf\{t : X_t(\omega) < \varepsilon\}$ are Borel for $\varepsilon > 0$, but their limit as $\varepsilon \downarrow 0$ isn't D.

σ-fields associated with a stopping time

DEFINITION. Let T be a stopping time of a filtration $(\mathcal{F}_t^o)$ on Ω. The σ-field of events prior to T, denoted $\mathcal{F}_T^o$, consists of all events $A \in \mathcal{F}_\infty^o$ such that 52

(52.1) for all t, $A \cap \{T \leq t\}$ belongs to $\mathcal{F}_t^o$.

When T is a positive constant r, $\mathcal{F}_T^o$ is the σ-field $\mathcal{F}_r^o$ of the filtration ; hence the notation and name are reasonable. We shall not attempt to give an intuitive jusitification of (52.1) - the justifications will not be lacking later.

Let us set $\mathcal{G}_t^o = \mathcal{F}_{t+}^o$ for all t $(\mathcal{G}_{0-}^o = \mathcal{F}_{0-}^o,\ \mathcal{G}_\infty^o = \mathcal{F}_\infty^o)$ and let T be a stopping time of $(\mathcal{G}_t^o)$ that is, a "wide sense stopping time" of $(\mathcal{F}_t^o)$ (49). Then an event A belongs to $\mathcal{G}_T^o$ if and only if it belongs to $\mathcal{G}_\infty^o = \mathcal{F}_\infty^o$ and

(52.2) for all t, $A \cap \{T < t\}$ belongs to $\mathcal{F}_t^o$.

It is quite natural to denote this σ-field $\mathcal{G}_T^o$ by the notation $\mathcal{F}_{T+}^o$.

The following proposition is a mere reformulation of 52. However we introduce an operation on stopping times and some notation, which we will use often later on. 53

THEOREM. Let T be optional relative to $(\mathcal{F}_t^o)$. Then A belongs to $\mathcal{F}_T^o$, if and only if A belongs to $\mathcal{F}_\infty^o$ and the random variable T_A defined by

(53.1) $T_A(\omega) = T(\omega)$ if $\omega \in A$, $T_A(\omega) = +\infty$ if $\omega \notin A$

is optional.

We now wish to define, for every stopping time, a σ-field $\mathcal{F}_{T-}^o$. It would be tempting to introduce, as in the remark of no. 52, the family of σ-fields $\mathcal{H}_t^o = \mathcal{F}_{t-}^o$ and to set $\mathcal{F}_{T-}^o = \mathcal{H}_T^o$ for every stopping time T of this family. This definition would be useless : it happens frequently that $(\mathcal{F}_t)$ satisfies the usual conditions and further $\mathcal{F}_t = \mathcal{F}_{t-}$ for all t. The above definition would then lead us to $\mathcal{F}_T = \mathcal{F}_{T-}$ for every stopping time T, while the distinction between "past" and "strict past" turns out to be important, even for stopping times of such families.

The correct definition has been given by Chung and Doob :

DEFINITION. Let T be a stopping time of $(\mathcal{F}_t^o)$. The σ-field of events strictly prior to T, denoted $\mathcal{F}_{T-}^o$, is the σ-field generated by $\mathcal{F}_{0-}^o$ and the events of the form 54

(54.1) $A \cap \{t < T\}$, $t \geq 0$, $A \in \mathcal{F}_t^o$.

The reader will verify that $\mathcal{F}_{T-}^o$ is also generated by the following (less convenient) sets

(54.2) $A \cap \{t \leq T\}$, $t \geq 0$, $A \in \mathcal{F}_{t-}^o$.

This definition is meaningful for every r.v. $T \geq 0$ (cf. (68.1)). If we write $\mathcal{G}_t^o = \mathcal{F}_{t+}^o$ for all t, $\mathcal{G}_{0-}^o = \mathcal{F}_{0-}^o$, then $\mathcal{G}_{T-}^o = \mathcal{F}_{T-}^o$ for all T.

ELEMENTARY PROPERTIES OF STOPPING TIMES

In the statements which follow, the r.v. are all defined on the same space $(\Omega,\mathcal{F}^o,\mathbb{P})$ and the stopping times relative to the same filtration $(\mathcal{F}^o_t)$, unless otherwise mentioned.

55 THEOREM. (closure properties).

(a) Let S and T be two stopping times ; then $S \wedge T$ and $S \vee T$ are stopping times.

(b) Let (S_n) be an increasing sequence of stopping times. Then $S = \lim_n S_n$ is a stopping time.

(c) Let (S_n) be a decreasing sequence of stopping times. Then $S = \lim_n S_n$ is a stopping time of the family $(\mathcal{F}^o_{t+})$ - a stopping time of $(\mathcal{F}^o_t)$ if the sequence is stationary, i.e. if for all ω there exists an integer n such that $S_m(\omega) = S_n(\omega)$ for all $m \geq n$.

Proof : (a) $\{S \wedge T \leq t\} = \{S \leq t\} \cup \{T \leq t\}$ belongs to $\mathcal{F}^o_t$; similarly for $\vee$.
(b) $\{S \leq t\} = \bigcap_n \{S_n \leq t\}$ belongs to $\mathcal{F}^o_t$. (c) $\{S < t\} = \bigcup_n \{S_n < t\}$ belongs to $\mathcal{F}^o_t$; when the sequence is stationary then also $\{S \leq t\} = \bigcup_n \{S_n \leq t\}$.

We deduce immediate consequences : the set of stopping times is closed under $(\vee c)$ and the set of stopping times of $(\mathcal{F}^o_{t+})$ (wide sense stopping times) is closed under the operations $(\vee c, \wedge c)$, and also under countable lim inf and lim sup.

56 THEOREM. (events prior to stopping times).

(a) For every stopping time S, $\mathcal{F}^o_{S-} \subset \mathcal{F}^o_S$ and S is $\mathcal{F}^o_{S-}$-measurable.

(b) Let S and T be two stopping times such that $S \leq T$. Then $\mathcal{F}^o_S \subset \mathcal{F}^o_T$, $\mathcal{F}^o_{S-} \subset \mathcal{F}^o_{T-}$ and, if $S < T$ everywhere, $\mathcal{F}^o_S \subset \mathcal{F}^o_{T-}$.

(c) Let S and T be two stopping times. Then

(56.1) for all $A \in \mathcal{F}^o_S$, $A \cap \{S \leq T\}$ belongs to $\mathcal{F}^o_T$

(56.2) for all $A \in \mathcal{F}^o_S$, $A \cap \{S < T\}$ belongs to $\mathcal{F}^o_{T-}$.

In particular,

(56.3) $\{S \leq T\}$, $\{S = T\}$ belong to $\mathcal{F}^o_S$ and $\mathcal{F}^o_T$ and $\{S < T\}$ to $\mathcal{F}^o_S$ and $\mathcal{F}^o_{T-}$.

(d) Let (S_n) be an increasing (resp. decreasing) sequence of stopping times and $S = \lim_n S_n$. Then $\mathcal{F}^o_{S-} = \bigvee_n \mathcal{F}^o_{S_n-}$ (resp. $\mathcal{F}^o_{S+} = \bigcap_n \mathcal{F}^o_{S_n+}$).

(e) Let S be a stopping time and $A \subset \Omega$. If $A \in \mathcal{F}^o_{\infty-}$ (resp. $A \in \mathcal{F}^o_\infty$), the set $A \cap \{S = \infty\}$ belongs to $\mathcal{F}^o_{S-}$ (resp. $\mathcal{F}^o_S$).

Proof : (a) We must prove that every generator B of $\mathcal{F}^o_{S-}$ belongs to $\mathcal{F}^o_S$, i.e. satisfies $B \cap \{S \leq t\} \in \mathcal{F}^o_t$ for all t. We take either $B \in \mathcal{F}^o_{0-}$ or $B = A \cap \{r < S\}$, $r \geq 0$, $A \in \mathcal{F}^o_r$. The verification is obvious.

For the second assertion, we note that for all t the set $\{S > t\}$ appears among the generators (54.1) of $\mathcal{F}^o_{S-}$.

(c) Property (56.1) follows from the following equality, true for all t :

$$A \cap \{S \leq T\} \cap \{T \leq t\} = [A \cap \{S \leq t\}] \cap \{T \leq t\} \cap \{S \wedge t \leq T \wedge t\}.$$

If A is $\mathcal{F}^0_S$-measurable, the three events appearing on the right are in $\mathcal{F}^0_t$. Property (56.2) follows from the following equality, where r runs through the rationals :

$$A \cap \{S < T\} = \bigcup_r (A \cap \{S < r\}) \cap \{r < T\}.$$

If A belongs to $\mathcal{F}^0_S$, the events appearing on the right belong to the generating system (54.1) of $\mathcal{F}^0_{T-}$.

(b) If $S \leq T$ and $A \in \mathcal{F}^0_S$, then $A = A \cap \{S \leq T\} \in \mathcal{F}^0_T$; if $S < T$ everywhere, $A = A \cap \{S < T\}$ belongs to $\mathcal{F}^0_{T-}$. Finally, it suffices to verify that the generators (54.1) of $\mathcal{F}^0_{S-}$ belong to $\mathcal{F}^0_{T-}$ if $S \leq T$; but, if $B \in \mathcal{F}^0_t$, then

$$B \cap \{t < S\} = (B \cap \{t < S\}) \cap \{t < T\} \text{ and } B \cap \{t < S\} \in \mathcal{F}^0_t.$$

(d) In the case of an increasing sequence, $\bigcup_n \mathcal{F}^0_{S_n-} \subset \mathcal{F}^0_{S-}$ by (b). On the other hand, every generator (54.1) of the σ-field $\mathcal{F}^0_{S-}$ can be written as $A \cap \{t < S\} = \bigcup_n A \cap \{t < S_n\}$ with $t \geq 0$, $A \in \mathcal{F}^0_t$, and we see that it belongs to $\bigcup_n \mathcal{F}^0_{S_n-}$. In the case of a decreasing sequence, we recall that $\mathcal{F}^0_{S+}$ is the set of $A \in \mathcal{F}^0_\infty$ such that $A \cap \{S < t\} \in \mathcal{F}^0_t$ for all t. As $A \cap \{S < t\} = \bigcap_n A \cap \{S_n < t\}$, we see that $\mathcal{F}^0_{S+} \supset \bigcap_n \mathcal{F}^0_{S_n+}$ and the converse inclusion follows from (b).

(e) The set of all $A \in \mathcal{F}^0_{\infty-}$ such that $A \cap \{S = \infty\}$ belongs to $\mathcal{F}^0_{S-}$ is a σ-field. Hence it suffices to verify that it contains $\mathcal{F}^0_t$ for all t. Now if A belongs to $\mathcal{F}^0_t$, $A \cap \{n < T\}$ is a generator (54.1) of $\mathcal{F}^0_{S-}$ for all $n \geq t$ and hence $A \cap \{S = \infty\}$ belongs to $\mathcal{F}^0_{S-}$. Finally, the case of $\mathcal{F}^0_S$ is trivial.

REMARK. (a) The reader may begin to perceive one of the great principles of this theory : to extend to stopping times T and σ-fields $\mathcal{F}^0_T$ all that is known for constant times t and σ-fields $\mathcal{F}^0_t$. Thus (a) and (b) are the extension to stopping times of the monotonicity of $(\mathcal{F}^0_t)$; (c) is the extension to arbitrary pairs of stopping times of properties (52.2) and (54.1), relating to pairs consisting of a stopping time and a constant, and (d) is the extension to stopping times of the continuity properties of the families $(\mathcal{F}^0_{t-})$ and $(\mathcal{F}^0_{t+})$.

(b) <u>It is not true</u> in general that for all $A \in \mathcal{F}^0_{S-}$, $A \cap \{S \leq T\}$ belongs to $\mathcal{F}^0_{T-}$. We shall see in no. 72, (c) the correct extension of (54.2).

(c) Let S and T be two stopping times. Then we have

$$\mathcal{F}_{S \wedge T} = \mathcal{F}_S \cap \mathcal{F}_T \text{ , } \mathcal{F}_{S \vee T} = \mathcal{F}_S \vee \mathcal{F}_T$$

Indeed, if A belongs to $\mathcal{F}_S$ and to $\mathcal{F}_T$, $A \cap \{S \leq S \wedge T\}$ and $A \cap \{T \leq S \wedge T\}$ belong to $\mathcal{F}_{S \wedge T}$ according to (56.1) ; taking, their union, A belongs to $\mathcal{F}_{S \wedge T}$, so that $\mathcal{F}_S \cap \mathcal{F}_T \subset \mathcal{F}_{S \wedge T}$. The reverse inclusion is obvious. Similarly, if A belongs to $\mathcal{F}_{S \vee T}$, $A \cap \{S \vee T \leq S\}$ and $A \cap \{S \vee T \leq T\}$ belong to $\mathcal{F}_S \vee \mathcal{F}_T$; taking unions the same is true for A. The reader may show also that $\mathcal{F}_{(S \vee T)-} = \mathcal{F}_{S-} \vee \mathcal{F}_{T-}$, and that $\mathcal{F}_{(S \wedge T)-} = \mathcal{F}_{S-} \cap \mathcal{F}_{T-}$ if S and T are predictable.

THEOREM. (a) <u>Let</u> S <u>be a stopping time of the family</u> $(\mathcal{F}^0_t)$ <u>and</u> T <u>be an</u> $\mathcal{F}^0_S$<u>-measurable</u> 57
<u>r.v. such that</u> $S \leq T$. <u>Then</u> T <u>is a stopping time of</u> $(\mathcal{F}^0_t)$. <u>The same conclusion holds</u> <u>if</u> S <u>is a stopping time of</u> $(\mathcal{F}^0_{t+})$, T <u>is</u> $\mathcal{F}^0_{S+}$<u>-measurable</u> $S < T$ <u>on</u> $\{S < \infty\}$.

This applies in particular to the r.v. $T = S + t$ $(t > 0)$ and to

(57.1) $$T = S^{(n)} = \sum_{k \geq 1} k2^{-n} I_{\{(k-1)2^{-n} \leq S \leq k2^{-n}\}}.$$

(b) Suppose that the family $(\mathcal{F}_t^o)$ is right continuous. Let S be a stopping time, $(\mathcal{G}_t^o)$ the family $(\mathcal{F}_{S+t}^o)$ and T and $\mathcal{F}_\infty^o$-measurable positive r.v. Then $U = S + T$ is a stopping time of $(\mathcal{F}_t^o)$, if and only if T is a stopping time of $(\mathcal{G}_t^o)$ (1).

Proof : (a) For all u

$$\{T \leq u\} = \{T \leq u\} \cap \{S \leq u\}.$$

As $\{T \leq u\}$ belongs to $\mathcal{F}_S^o$, this belongs to $\mathcal{F}_u^o$ by definition of $\mathcal{F}_S^o$, and T is a stopping time. If $S < T$ on $\{S < \infty\}$, we can replace $\{S \leq u\}$ by $\{S < u\}$ in the argument ; details are left to the reader.

(b) Suppose that T is a stopping time of $(\mathcal{G}_t^o)$. We write that

$$\{U < t\} = \bigcup_b \{S + b < t\} \cap \{T < b\}$$

on all rationals $b < t$. But $\{T < b\} \in \mathcal{G}_b^o = \mathcal{F}_{S+b}^o$, hence, by definition of $\mathcal{F}_{S+b}^o$, $\{T < b\} \cap \{S + b < t\} \in \mathcal{F}_t^o$ and $\{U < t\} \in \mathcal{F}_t^o$. Since the family $(\mathcal{F}_t^o)$ is right continuous, U is a stopping time of it. Conversely, we suppose for simplicity that S is finite. Then if U is a stopping time of $(\mathcal{F}_t^o)$

$$\{T \leq t\} = \{S + T \leq S + t\} \in \mathcal{F}_{S+t}^o = \mathcal{G}_t^o \quad \text{by (56.1)}$$

and T is indeed a stopping time of $(\mathcal{G}_t^o)$.

58 COROLLARY. Every stopping time S of the family $(\mathcal{F}_{t+}^o)$ is the limit of a decreasing sequence of discrete stopping times of the family $(\mathcal{F}_t^o)$ and can also be represented as the lower envelope of an (in general non-decreasing) sequence of stopping times T of the following type

(58.1) $$T = a.I_A + (+\infty)I_{A^c}, \quad a \in \mathbb{R}_+, \ A \in \mathcal{F}_a^o.$$

Proof : We have $S = \lim_n S^{(n)} = \inf_n S^{(n)}$ (57.1) and $S^{(n)}$ is the infumum of the stopping times

$$k.2^{-n} I_{A^c} + (+\infty) I_{A_k^c} \text{ where } A_k = \{S^{(n)} = k.2^{-n}\}.$$

We immediately give an important application to the completion of σ-fields (due to Dynkin).

59 THEOREM. Let $\mathcal{H}_t$ be the σ-field obtained by adjoining to $\mathcal{F}_t^o$ all the $\mathbb{P}$-negligible sets. Let T be a stopping time of the family $(\mathcal{H}_t)$. There then exists a stopping time U of the family $(\mathcal{F}_{t+}^o)$ such that $T = U$ $\mathbb{P}$-a.s. Further, for all $L \in \mathcal{H}_T$, there exists $M \in \mathcal{F}_{U+}^o$ such that $L = M$ $\mathbb{P}$-a.s.

Proof : For the first assertion it suffices to treat the case of a stopping time of

(1) A more complete result appears in the first edition, no. 57.

type (58.1) and then pass to the lower envelope. We then write $U = a.I_B + (+\infty)I_{B^c}$, where B is an element of $\mathcal{F}^o_a$ $\mathbb{P}$-a.s. equal to $A \in \mathcal{H}_a$ (passing to the limit is responsible for the appearance of the family $(\mathcal{F}^o_{t+})$ instead of $(\mathcal{F}^o_t)$).

The second assertion is a consequence of the first one. Since $L \in \mathcal{H}_\infty$, we choose some $L' \in \mathcal{F}^o_\infty$ such that $L' = \mathbb{P}$-a.s. and then a stopping time V of the family $(\mathcal{F}^o_{t+})$ such that $V = T_L$ a.s. The required event M then is $(L' \cap \{U = \infty\}) \cup \{V = U < \infty\}$.

Stochastic intervals. Optional and predictable σ-fields.

DEFINITION. Let U and V be two positive real valued functions on Ω such that U ≤ V. We denote by $[\![U,V[\![$ the subset of $\mathbb{R}_+ \times \Omega$ 60

(60.1) $$[\![U,V[\![= \{(t,\omega) : U(\omega) \le t < V(\omega)\}.$$

We define similarly the "stochastic intervals" of types $]\!]U,V]\!]$, $]\!]U,V[\![$ and $[\![U,V]\!]$. We set in particular $[\![U,U]\!] = [\![U]\!]$ (the graph of U).

The use of double brackets $[\![$ aims at distinguishing [1], when U and V are constant stopping times, the stochastic interval $[\![U,V[\![\subset \mathbb{R}_+ \times \Omega$ from the ordinary interval $[U,V[$ of $\mathbb{R}_+$. Note that we are concerned with subsets of $\mathbb{R}_+ \times \Omega$ and not $\bar{\mathbb{R}}_+ \times \Omega$.

Here is a fundamental definition relating to a filtration $(\mathcal{F}^o_t)$.

DEFINITION [2]. The σ-field $\mathcal{O}$ on $\mathbb{R}_+ \times \Omega$ generated by the real valued processes $(X_t)_{t \ge 0}$, adapted to the family $(\mathcal{F}^o_t)$ and with r.c.l.l. paths (16) is called the optional or well measurable σ-field. 61

The σ-field $\mathcal{P}$ on $\mathbb{R}_+ \times \Omega$ generated by the processes $(X_t)_{t \ge 0}$, adapted to the family $(\mathcal{F}^o_{t-})$ [3] and with left-continuous paths on $]0,\infty[$ is called the predictable σ-field.

REMARKS. (a) If we deal with $\bar{\mathbb{R}}_+ \times \Omega$, the processes in this definition must include an additional variable X_∞. In the optional case we require that it be $\mathcal{F}^o_\infty$-measurable ; in the predictable case we require left-continuity at infinity.

(b) In a deeper study of the theory of processes, one realizes that it would be more natural to consider the two σ-fields as defined on different spaces : the optional σ-field on $[0,\infty[\times \Omega$, but the predictable σ-field on $]0,\infty] \times \Omega$. In all considerations on the predictable σ-field, time 0 plays the devil's role.

(c) If we replace the filtration $(\mathcal{F}^o_t)$ by the right continuous filtration $\mathcal{G}^o_t = \mathcal{F}^o_{t+}$, $\mathcal{G}^o_{0-} = \mathcal{F}^o_{0-}$, the optional σ-field is enlarged but the predictable σ-field does not change.

(1) In many papers, authors are less formal about it and use ordinary brackets.

(2) We shall comment in no. 68 on the slight lack of symmetry between the two parts of the definition.

(3) By left-continuity, this is equivalent - except at 0 - to adaptation to $(\mathcal{F}^o_t)$.

(d) A process (X_t) defined on Ω is said to be _optional_ (resp. _predictable_) if the function $(t,\omega)\mapsto X_t(\omega)$ on $\mathbb{R}_+\times\Omega$ is measurable given the optional (resp. predictable) σ-field.

62 EXAMPLES [1] (a) All the stochastic intervals determined by a pair (S,T) of stopping times such that $S\le T$ are optional sets: The stochastic intervals $[0,T[$, $[S,\infty[$ indeed have adapted indicators with r.c.l.l. paths and the same also holds for their intersection $[\![S,T[\![$. Taking $T = S + 1/n$ and letting n tend to infinity, we see that $[\![S]\!]$ is optional and then we deduce also the optionality of all other stochastic intervals.

Let Z be an $\mathcal{F}^0_S$-measurable r.v. and let $Z.I_{[\![S,T[\![}$ denote the process (X_t) defined by $X_t(\omega) = Z(\omega).I_{[\![S,T[\![}(t,\omega)$. It is easily verified that $Z.I_{[\![S,T[\![}$ is optional : approximating Z by $\mathcal{F}^0_S$-measurable elementary r.v., we are immediately reduced to the case where Z is the indicator of an element A of $\mathcal{F}^0_S$ and then $Z.I_{[\![S,T[\![}$ is the indicator of $[\![S_A,T_A[\![$. There are analogous results for the other stochastic intervals.

(b) If S and T are two stopping times of $(\mathcal{F}^0_{t+})$ such that $S\le T$, the stochastic interval $]\!]S,T]\!]$ is predictable : for its indicator, the product of the indicators of $]\!]0,T]\!]$ and $]\!]S,\infty[\![$, is adapted to $(\mathcal{F}^0_{t-})$ and has left-continuous paths.

If Z is an $\mathcal{F}^0_{S+}$-measurable r.v., it is easily verified, as above, that the process $Z.I_{]\!]S,T]\!]}$ is predictable.

The following notation will be in constant use. To simplify, we give it only for real valued processes.

63 DEFINITION. _Let_ $(X_t)_{t\in\mathbb{R}_+}$ _be a real valued process on_ Ω _and let_ H _be a function on_ Ω _with values in_ $\bar{\mathbb{R}}_+$. _Then we denote by_ X_H _the function defined on_ $\{H<\infty\}$ _by_

(63.1) $X_H(\omega) = X_{H(\omega)}(\omega)$ ("state of the process X at time H")

and by $X_H I_{\{H<\infty\}}$ _the function equal to_ X_H _on_ $\{H<\infty\}$ _and to_ 0 _on_ $H=\infty\}$.

Of course, for a process $(X_t)_{t\in\bar{\mathbb{R}}_+}$ we can define X_H on the whole of Ω, and not only on $\{H<\infty\}$.

We begin to study the optional and predictable σ-fields.

64 THEOREM. (a) _The optional_ σ-_field is contained in the progressive_ σ-_field_ (the inclusion is usually stict) [2].

(b) _Let_ T _be a stopping time of_ $(\mathcal{F}^0_t)$ _and let_ (X_t) _be a progressive process (in particular, an optional process). Then the function_ $X_T I_{\{T<\infty\}}$ is $\mathcal{F}^0_T$_-measurable Conversely, if_ Y _is an_ $\mathcal{F}^0_T$_-measurable r.v., there exists an optional process_ $(X_t)_{t\in\bar{\mathbb{R}}_+}$ _such that_ $Y = X_T$.

---

(1) We shall see other examples later.

(2) It is strict, for instance, in the case of the natural filtration of Brownian motion.

(c) The optional σ-field is generated by the stochastic intervals $[\![S,\infty[\![$, where S is a stopping time.

Proof : (a) has already been proved (15).

(b) Let (X_s) be a progressive process ; for all t, the mapping $(s,\omega)\mapsto X_s(\omega)$ is $\mathcal{B}([0,t])\times\mathcal{F}^o_t$-measurable on $[0,t]\times\Omega$, and the mapping $\omega\mapsto T(\omega)\wedge t$ is $\mathcal{F}^o_t$-measurable since T is a stopping time. By composition, we see that $\omega\mapsto X_{T(\omega)\wedge t}(\omega)$ is $\mathcal{F}^o_t$-measurable. Then let $Y = X_T I_{\{T<\infty\}}$; for all t, $YI_{\{T\le t\}} = X_{T\wedge t}I_{\{T\le t\}}$ is $\mathcal{F}^o_t$-measurable and this means that Y is $\mathcal{F}^o_T$-measurable (52.1).

The converse is obvious : if Y is $\mathcal{F}^o_T$-measurable, the process (X_t) defined for $0\le t\le\infty$ by $X_t = YI_{\{t\ge T\}}$ is adapted with r.c.l.l. paths, and $Y = X_T$.

(c) Let $\mathcal{J}_0$ be the paving consisting of the stochastic intervals $[S,T[$, where S and T are stopping times and $S\le T$. Then $\mathcal{J}_0\subset\mathcal{O}$ and we must examine whether every process (X_t) adapted to $(\mathcal{F}^o_t)$ and with r.c.l.l. paths is $\sigma(\mathcal{J}_0)$-measurable. We choose a number $\varepsilon>0$ and introduce inductively the following functions (where d is a metric on $\overline{\mathbb{R}}$)

(64.1)
$$T^\varepsilon_0 = 0,\ T^\varepsilon_{n+1}(\omega) = \inf\{t > T^\varepsilon_n(\omega) : d(X_{T^\varepsilon_n(\omega)}, X_t(\omega)) \ge \varepsilon \text{ or } d(X_{T^\varepsilon_n(\omega)}, X_{t-}(\omega)) \ge \varepsilon\}$$

(64.2)
$$Z^\varepsilon_n(\omega) = X_{T^\varepsilon_n}(\omega) I_{\{T^\varepsilon_n<\infty\}}(\omega).$$

The path deviates from $Z^\varepsilon_n(\omega)$ by less than ε on the interval $[T^\varepsilon_n(\omega), T^\varepsilon_{n+1}(\omega)[$, but its oscillation on the closed interval $[T^\varepsilon_n(\omega), T^\varepsilon_{n+1}(\omega)]$ is $\ge\varepsilon$ if $T^\varepsilon_{n+1}(\omega)<\infty$. Hence the existence of left-hand limits prevents the T^ε_n from accumulating at a finite distance. Thus (X_t) is the uniform limit, as $\varepsilon\to 0$, of the processes

(64.3)
$$X^\varepsilon_t(\omega) = \sum_0^\infty Z^\varepsilon_n(\omega) I_{[T^\varepsilon_n, T^\varepsilon_{n+1}[}(t,\omega)$$

and it suffices to show that (X_t) is $\sigma(\mathcal{J}_0)$-measurable. We first verify that the T_n are stopping times of the family $(\mathcal{F}^o_t)$: to simplify the notation, we do this only for the first of them and leave to the reader the easy induction on n. We then write

(64.4)
$$T = \inf\{t>0 : d(X_0,X_t) \text{ or } d(X_0,X_{t-}) \ge \varepsilon\}$$

and check that $\{T\le t\}\in\mathcal{F}^o_t$. To this end we remark that the set between { }is closed, hence the relation $T = t$ implies that $d(X_0,X_t)\ge\varepsilon$ or $d(X_0,X_{t-})\ge\varepsilon$ and the relation $\{T\le t\}$ is equivalent to the following, where $\mathbb{Q}_t$ denotes the set consisting of the rationals of $]0,t[$ and the point t :

$$\forall n\in\mathbb{N}\ \exists r_n\in\mathbb{Q}_t,\ d(X_0,X_{r_n}) > \varepsilon - 1/n.$$

Since $\mathbb{Q}_t$ is countable, clearly this set belongs to $\mathcal{F}^o_t$.

Having established this point concerning the T_n^ε, it remains to remark (leaving out the useless ε and n) that processes of the form $ZI_{[S,T[}$, where S and T are two stopping times and Z is $\mathcal{F}_S^o$-measurable, are $\sigma(\mathcal{J}_0)$-measurable. If the reader has paid attention to Example 62, (a), he will find this obvious.

65 THEOREM. Under the usual conditions, the optional σ-field is also generated by the right-continuous processes adapted to the family $(\mathcal{F}_t^o)$.

Proof : We show first that every right-continuous adapted process (X_t) is indistinguishable from an optional process. We return to a variant of the construction (64.1), but this time arguing by transfinite induction, since now left-hand limits may not exist, and therefore accumulation of stopping times can happen at a finite distance : we set $T_0 = 0$ and for every countable ordinal α

$$T_{\alpha+1}^\varepsilon = \inf\{t > T_\alpha^\varepsilon : d(X_{T_\alpha^\varepsilon}, X_t) > \varepsilon\},$$

on the other hand, for every limit ordinal β,

$$T_\beta^\varepsilon = \sup_{\alpha<\beta} T_\alpha^\varepsilon .$$

Z_α^ε, X^ε are defined as in (64.2), (64.3), but with the obvious modifications : α instead of n, with the sum over the ordinals. The T_α^ε are - thanks to the strict inequality "$>\varepsilon$" - stopping times of the family $(\mathcal{F}_{t+}^o)$, and the process (X_t) is approximated uniformly by the (X_t^ε). On the other hand, there exists by 0.8 an ordinal α such that $T_\alpha^\varepsilon = +\infty$ $\mathbb{P}$-a.s. and the process (X_t^ε) is therefore P-indistinguishable from an optional process (Y_t). The process (Z_t) defined by $Z_t = X_t - Y_t$ then is adapted, right-continuous and evanescent. We now show that it is optional, which will complete the proof. We define, for n, a process (Z_t^n) by writing $Z_0^n = Z_0$ and

$$Z_t^n = Z_{(k+1)/n} \text{ for } t \in]k/n, (k+1)/n]$$

where k runs through the integers. Since (Z_t) is evanescent, $Z_{(k+1)/n}$ is $\mathbb{P}$-a.s. zero and hence $\mathcal{F}_{k/n}$-measurable (we are under the usual conditions). Then the process (Z_t^n) is optional and so is also its limit (Z_t) as n tends to infinity.

The proof of the theorem gives immediately a result on modifications of processes.

66 THEOREM. Let (X_t) be an optional process with respect to the family $(\mathcal{F}_t^o)$. Then there exists a predictable process (Y_t) such that the set

(66.1) $$A = \{(t,\omega) : X_t(\omega) \neq Y_t(\omega)\}$$

is the union of a sequence of graphs of stopping times. In particular, the section $A(\omega)$ is countable for all ω.

Proof : We note first that it suffices to prove that A is contained in the union of a sequence of graphs of stopping times : for if T is a stopping time, A being progressive, the event $L = \{\omega : (T(\omega),\omega) \in A\}$ belongs to $\mathcal{F}_T^o$ (64, (b)) and the stopping time T_L has its graph contained in A. If A is contained in a union of graphs $[T^n]$,

A then is equal to the union of the corresponding graphs $[\![T^n_{L_n}]\!]$, where $L_n = \{\omega : (T^n(\omega), \omega) \in A\}$.

On the other hand, we may reduce to bounded real valued processes. Then let $\mathcal{H}$ be the set of optional bounded real processes (X_t) for which the statement is true. By truncating, we may also assume that the corresponding predictable processes are bounded. We then verify that $\mathcal{H}$ is an algebra closed under monotone convergence (if processes $(X^n_t) \in \mathcal{H}$ increase to a bounded process (X_t) and predictable processes (Y^n_t) satisfy the statement relative to (X^n_t), then so does the process $Y_t = \liminf_n Y^n_t$ relative to (X_t)). By the monotone class theorem, it suffices to verify the statement for processes generating the optional σ-field. We choose for (X_t) the indicators of intervals $[\![S,T[\![$, where S and T are stopping times, the corresponding predictable processes (Y_t) being the indicators of the intervals $]\!]S,T]\!]$.

We now give the theorem analogous to 64, concerning the predictable σ-field.

THEOREM. (a) The predictable σ-field is contained in the optional σ-field (and hence in the progressive σ-field). 67

(b) Let T be a stopping time of the family $(\mathcal{F}^0_t)$ and let (X_t) be a predictable process. Then the function $X_T I_{\{T < \infty\}}$ is $\mathcal{F}^0_T$-measurable. Conversely, if Y is an $\mathcal{F}^0_T$-measurable r.v. there exists a predictable process $(X_t)_{t \in \bar{\mathbb{R}}_+}$ such that $Y = X_T$.

(c) The predictable σ-field is generated by the sets of the form

(67.1) $\{0\} \times A,\ A \in \mathcal{F}^0_{0-}$ and $]s,t] \times A\ (0 < s < t$, s and t rationals, $A \in \bigcup_{r<s} \mathcal{F}^0_r)$

or also the sets

(67.2) $\{0\} \times A,\ A \in \mathcal{F}^0_{0-}$ and $[s,t[\times A\ (0 < s < t$, s and t rationals, $A \in \bigcup_{r<s} \mathcal{F}^0_r)$.

Proof : Though the order of the statement is similar to that of 64, we now begin with the proof of (c).

We first note that, by 62, every stochastic interval $]\!]S,T]\!]$, where S and T are stopping times (even in the wide sense), is predictable. Conversely, let (X_t) be a process adapted to $(\mathcal{F}^0_{t-})$ with left-continuous paths ; X_t is the limit of the processes

(67.3) $$X^n = X_0 I_{[\![0]\!]} + \sum_{k \geq 0} X_{k2^{-n}} I_{]\!]k2^{-n},(k+1)2^{-n}]\!]}.$$

By representing $X_{k2^{-n}}$ as an increasing limit of $\mathcal{F}^0_{(k2^{-n})}$-measurable elementary r.v., we see that (X_t) is measurable with respect to the σ-field generated by the sets $\{0\} \times A (A \in \mathcal{F}^0_{0-})$ and the stochastic intervals $]\!]S,T]\!]$ of the following special type :

(67.4) $S = s$ on A, $+\infty$ on A^c, $T = t$ on A, $+\infty$ on A^c with $0 < s < t$, $A \in \mathcal{F}^0_{s-}$.

We can moreover replace the condition $A \in \mathcal{F}^0_{s-}$ by $A \in \bigcup_{r<s} \mathcal{F}^0_r$. Hence we get the generating system (67.1), and deduce from it (67.2)

(a) is obvious : the predictable σ-field is generated by stochastic intervals and these belong to the optional σ-field.

Finally we prove (b). To simplify some of the work, we suppose that T is finite. The statement then means that $\mathcal{F}^o_{T-}$ is the inverse image of the predictable σ-field $\mathcal{P}$ under the mapping $f : \omega \mapsto (T(\omega),\omega)$ of Ω into $\mathbb{R}_+ \times \Omega$.

Let first verify that f is measurable from $\mathcal{F}^o_{T-}$ into $\mathcal{P}$: it suffices to show that for sets U generating the predictable σ-field, $f^{-1}(U)$ belongs to $\mathcal{F}^o_{T-}$ and to thus end we choose $U = \{0\} \times A$ $(A \in \mathcal{F}^o_{0-})$, $U =]s,\infty[\times A$ $(A \in \mathcal{F}^o_s)$ whose inverse images are $A \cap \{T = 0\}$, $A \cap \{s < T\}$. These sets belongs to $\mathcal{F}^o_{T-}$: in the first case because T is $\mathcal{F}^o_{T-}$ -measurable and in the second case by Definition 54.

We now verify that every $A \in \mathcal{F}^o_{T-}$ is the inverse image of a predictable set, i.e. that $f^{-1}(\mathcal{P})$ contains the generators 54 of $\mathcal{F}^o_{T-}$ - this is immediate by (c) and we leave the details to the reader.

68 REMARKS. (a) Provided the σ-fields $\mathcal{F}^o_{t-}$ are separable, the predictable σ-field is separable this is seen immediately from the generating system (67.1) or (67.2).

(b) Given an <u>arbitrary</u> $\mathcal{F}$-measurable positive random variable L, it is natural to define the σ-fields associated with L as follows :

(68.1) <u>a random variable</u> Y <u>is measurable with respect to</u> $\mathcal{F}^o_L$ (<u>resp</u>. $\mathcal{F}^o_{L-}$) <u>if and only if there exists an optional</u> (<u>resp. predictable</u>) <u>process</u> $(X_t)_{t \in \overline{\mathbb{R}}_+}$ <u>such that</u> $Y = X_L$.

We define similarly $\mathcal{F}^o_{L+}$ by considering optional processes of the family $(\mathcal{F}^o_{t+})$. This definition is useful in more advanced parts of the theory

(c) The lack of symmetry between the two parts of definition 61 (r.c.l.l. process on the one hand and left-continuous on the other) is only apparent : the predictable σ-field is also generated by left-continuous processes <u>with right-hand limits</u>! But in fact the predictable σ-field is even generated by the adapted processes with <u>continuous</u> paths : for if A belongs to $\mathcal{F}^o_{0-}$, the process $X_t = tI_A + I_{A^c}$ is adapted (to the family $(\mathcal{F}^o_{t-})$) and has continuous paths and $\{0\} \times A = \{(t,\omega) : X_t(\omega) = 0\}$; similarly, if S is a stopping time, the process $X_t = (t-S)^+$ is adapted with continuous paths and the stochastic interval $]\!]S,\infty[\![$ is equal to $\{(t,\omega)\} : X_t(\omega) > 0$.

Predictable stopping times

69 The following definition is not the most usual one for predictable stopping times, but it is more convenient for arguments without probability and it is equivalent to the current definition under the "usual conditions" (see no. 77).

We have seen above that if T is a stopping time, the interval $[\![T,\infty[\![$ is optional. But conversely, if T is a positive function on Ω and the interval $[\![T,\infty[\![$ is optional, T is a stopping time - for if X denotes the indicator of $[\![T,\infty[\![$, X_t is $\mathcal{F}^0_t$-measurable (64, (b)) and it is the indicator of $\{T \leq t\}$. Replacing the optional σ-field by the predictable σ-field, we may define :

DEFINITION. *Let* T *be a mapping of* Ω *into* $\bar{\mathbb{R}}_+$. *We say that* T *is a* predictable[1] r.v. *or a* predictable stopping time *(of the family* $(\mathcal{F}^0_t)$*) if the stochastic interval* $[\![T,\infty[\![$ *is predictable.*

Since $\mathcal{P} \subset \mathcal{O}$, a predictable r.v. is a stopping time, whence the terminology. Constants obviously are predictable stopping times.

The discrete case may help in understanding this definition. We are given a family $(\mathcal{F}^0_n)_{n \geq -1}$, $\mathcal{F}^0_{-1}$ playing the role of $\mathcal{F}^0_{0-}$ in the continuous case. A predictable process then is a process $(X_n)_{n \geq 0}$ such that, for all $n \geq 0$, X_n is $\mathcal{F}^0_{n-1}$-measurable and a predictable stopping time is a stopping time T such that, for all $n \geq 0$ the set $\{T\} = n$ belongs to $\mathcal{F}^0_{n-1}$.

If we replace the family $(\mathcal{F}^0_t)$ by the family $\mathcal{G}^0_t = \mathcal{F}^0_{t+}$, $\mathcal{G}^0_{0-} = \mathcal{F}^0_{0-}$, we do not change the σ-field $\mathcal{P}$ so the predictable stopping times remain the same. The same is true if we replace $\mathcal{F}^0_t$ by $\mathcal{F}^0_{t-}$.

We now connect this definition with the existence of a sequence of "precursory signs" for the stopping time T.

DEFINITION. *Let* T *be a stopping time of the family* $(\mathcal{F}^0_t)$ *and* (T_n) *be an increasing* 70
sequence of stopping times (may be only in the wide sense) such that $T_n \leq T$ *for all* n. *We say that the sequence* (T_n) foretells T on $A \subset \Omega$ *if on the set* A

(70.1) $\qquad \lim_n T_n = T,\ T_n < T$ *for all* n.

If these properties hold $\mathbb{P}$-a.s. on A, we of course say that the sequence *foretells*[2] T *a.s.* on A. Since the condition $T_n \leq T$ implies that $T_n = T$ on $\{T = 0\}$, we say briefly that the sequence T_n *foretells* T (foretells T a.s.) without mentioning any set A, to mean that (T_n) foretells T (a.s.) *on the set* $\{T > 0\}$. We say that T is *foretellable*[3] ($\mathbb{P}$-foretellable) if there exists a sequence (T_n) of stopping times in the wide sense foretelling T (foretelling T $\mathbb{P}$-a.s.).

Under the usual conditions, every $\mathbb{P}$-foretellable stopping time is foretellable. For let (T_n) be a sequence foretelling T on A^c, where A has measure zero ; then A belongs to $\mathcal{F}_0$ and the stopping times $T'_n = T_n I_{A^c} + (T-1/n)^+ I_A$ foretell T everywhere.

We may now give the fundamental example of predictable stopping times. We shall see later that there is an "almost-converse" : every predictable stopping time is $\mathbb{P}$-foretellable for every law $\mathbb{P}$.

(1) The word *previsible* is sometimes used with the same meaning.
(2) Some authors say "to announce" instead of "to foretell".
(3) This terminology is used only in the few following numbers.

71 THEOREM. (a) Let T be a foretellable stopping time such that $\{T = 0\}$ belongs to $\mathcal{F}^o_{0-}$; [1] then T is predictable.

(b) Every $\mathbb{P}$-foretellable stopping time T is a.s. equal to a foretellable stopping time, and hence to a predictable stopping time if $\mathcal{F}^o_0 = \mathcal{F}^o_{0-}$.

Proof : (a) Let (T_n) be a sequence of stopping times in the wide sense foretelling T. Then the interval $[\![T,\infty[\![$ is the union of the predictable sets $\{0\} \times \{T = 0\}$ and $]\!]T_n,\infty[\![$ and is hence predictable.

To prove (b), let (T^n) be a sequence which foretells T a.s. We set $T' = \lim_n T_n$ and, for all n, $B_n = \{T = 0\} \cup \{T_n < T'\}$, which belongs to $\mathcal{F}^o_{T^n+}$ (56.3). Then the wide sense stopping times $S^n = T^n_{B_n} \wedge n$ are increasing, their limit S is equal to T a.s. and the sequence (S^n) obviously foretells S.

72 Here is a simple example : let S be a wide sense stopping time and T be a $\mathcal{F}^o_{S+}$-measurable r.v. such that $S < T$ on $\{S < \infty\}$; then T is a stopping time (57) : since the sequence $T_n = n \wedge (\frac{1}{n}S + \frac{n-1}{n}T)$ foretells T, T is a predictable stopping time.

> The identity - to within sets of measure zero - between the notions of predictable stopping times and foretellable stopping times is the origin of the word "predictable" and gives it intuitive content. We have mentioned earlier that the idea of a stopping time is that of the "first time some given random phenomenon occurs". The existence of a foretelling sequence means that this phenomenon cannot take use by surprise : we are forewarned, by a succession of precursery signs, of the exact time the phenomenon will occur.

In the following statement we group the elementary properties of predictable stopping times.

73 THEOREM. (a) The set of predictable stopping times is closed under the operations $\wedge$ and $\vee$.

(b) Let S be a predictable stopping time and T a stopping time. Then

(73.1) if A belongs to $\mathcal{F}^o_{S-}$, $A \cap \{S \leq T\} \in \mathcal{F}^o_{T-}$.

In particular, the events $\{S < T\}$, $\{S = T\}$; $\{S \leq T\}$ belong to $\mathcal{F}^o_{T-}$.

(c) Let S be a predictable stopping time and A be an element of $\mathcal{F}^o_{\infty-}$. In order that $A \in \mathcal{F}^o_{S-}$, it is necessary and sufficient that S_A be predictable.

(d) Let (S_n) be an increasing sequence of predictable stopping times and let $S = \lim_n S_n$; then S is predictable. The same result is true for a stationary decreasing sequence (S_n) (i.e. such that for all ω there exists N such that $S_{N+k}(\omega) = S_N(\omega)$ for all k).

Proof : (a) We simply remark that $[\![S \vee T,\infty[\![= [\![S,\infty[\![\wedge [\![T,\infty[\![$ and $[\![S \wedge T,\infty[\![= [\![S,\infty[\![\cup [\![T,\infty[\![$.

---

(1) In the usual cases $\mathcal{F}^o_{0-} = \mathcal{F}^o_0$ and this condition is always satisfied.

(b) We already know that $A \cap \{S < T\} \in \mathcal{F}^0_{T-}$ (56.2) and $A \cap \{T = \infty\} \in \mathcal{F}^0_{T-}$ (56(e)). Hence it suffices to consider $A \cap \{S = T < \infty\}$. Let I be the indicator of this set ; let (X_t) be a predictable process such that $I_A = X_S I_{\{S < \infty\}}$ (67(b)) ; let J be the indicator of the predictable set $[\![S]\!] = [\![S,\infty[\![\setminus]\!]S,\infty[\![$. Then $I = X_T J_T I_{\{T < \infty\}}$ and we conclude using 67(b).

Taking $A = \Omega$, we get $\{S \leq T\} \in \mathcal{F}^0_{T-}$; we also know that $\{S < T\} \in \mathcal{F}^0_{T-}$ (56.2) and hence, taking a difference, $\{S = T\} \in \mathcal{F}^0_{T-}$.

(c) If the stopping time S_A is predictable, let (X_t) be the (predictable) indicator of $[\![S_A,\infty[\![$; we know that $X_S I_{\{S < \infty\}}$ is $\mathcal{F}^0_{S-}$-measurable (67(b)) and hence $A \cap \{S < \infty\} \in \mathcal{F}^0_{S-}$. Similarly for $A \cap \{S = \infty\}$ since $A \in \mathcal{F}^0_{\infty-}$ (56(e)).

Conversely, let $A \in \mathcal{F}^0_{S-}$; there exists (67(b)) a predictable process (X_t) such that I_A and X_S coincide on $\{S < \infty\}$. The graph $[\![S_A]\!]$ then is the intersection of $[\![S]\!]$ and of the predictable set $\{(t,\omega) = X_t(\omega) = 1\}$. The graph $[\![S]\!]$ is equal to $[\![S,\infty[\![\setminus]\!]S,\infty[\![$ and is hence predictable ; thus $[\![S_A]\!]$ is predictable and so is $[\![S_A,\infty[\![=]\!]S_A,\infty[\![\cup [\![S_A]\!]$.

(d) In the first case, $[\![S,\infty[\![= \bigcap_n [\![S_n,\infty[\![$ and in the second case $[\![S,\infty[\![= \bigcup_n [\![S_n,\infty[\![$.

REMARK. The predictable σ-field is generated by the intervals $[\![S,T[\![$, where S and T are predictable, or even both predictable and foretellable. For among the generators (67.2) those of the second kind are of the form $[\![S,T[\![$ with $S = s_A (A \in \bigcup_{r<s} \mathcal{F}^0_r)$, $T = t_A$, foretellable by stopping times of the form $n \wedge (s - \frac{1}{n})_A$, $n \wedge (t - \frac{1}{n})_A$ with n sufficiently large. As for the generators of the first kind, if $A \in \mathcal{F}^0_{0-}$, then $\{0\} \times A = \bigcap_n [\![0_A, 0_A + 1/n[\![$ and 0_A is foretellable by the sequence $n \wedge 0_A$. 74

Sequences foretelling a predictable time

Our first task in proving the "converse" to 71 consists in establishing properties analogous to 73, but now for $\mathbb{P}$-foretellable stopping times.

LEMMA. Let $\mathcal{U}$ be the set of stopping times of $(\mathcal{F}^0_t)$ which are foretellable a.s. by a sequence of wide sense stopping times. 75

(a) $\mathcal{U}$ is closed under the operations $\vee$ and $\wedge$ and $0 \in \mathcal{U}$, $+\infty \in \mathcal{U}$.

(b) The limit of an increasing sequence of elements of $\mathcal{U}$ belongs to $\mathcal{U}$.

(c) The limit of a stationary decreasing sequence of elements of $\mathcal{U}$ belongs to $\mathcal{U}$.

(d) If S and T belong to $\mathcal{U}$, so does S_A, where $A = \{S < T\}$.

(e) Let $S \in \mathcal{U}$ and T be a stopping time such that $S = T$ a.s. ; then $T \in \mathcal{U}$.

Proof : (a) Let S and T be two elements of $\mathcal{U}$ which are foretold a.s. by sequences (S_n) and (T_n) ; then $S \wedge T$ and $S \vee T$ are foretold a.s. by the sequences $(S_n \wedge T_n)$ and $(S_n \vee T_n)$.

(b) Let (T_n) be an increasing sequence of elements of $\mathfrak{U}$ and for each n let $(T_n^p)_{p \in \mathbb{N}}$ be a sequence foretelling a.s. (T_n). Then the stopping time $T = \lim_n T_n$ is a.s. foretold by the sequence $T^p = T_1^p \vee \ldots \vee T_p^p$.

(c) Suppose now that the T_n, still a.s. foretold by the (T_n^p), form a stationary decreasing sequence and let $T = \lim_n T_n$. We consider some bounded metric d defining the topology on $\bar{\mathbb{R}}_+$ (for example $d(x,y) = |e^{-x} - e^{-y}|$) and choose for each n an integer n' such that $\mathbb{P}\{d(T_n^{n'},T_n) > 1/n\} \leq 2^{-n}$. Then we set

$$U_p = \inf_{n \geq p} T_n^{n'}.$$

The U^p are wide sense stopping times and the sequence (U^p) is increasing ; we denote its limit by U. Since $T_n^{n'} \leq T_n \leq T$, we have $U^p \leq T$ and hence $U \leq T$ everywhere. Since $T_n^{n'} < T_n$ a.s. on $\{T_n > 0\}$, $U^p < T_n$ a.s. on $\{T_n > 0\}$ for all $n \geq p$ and hence $U^p < T$ a.s. on $\{T > 0\}$, the sequence (T_n) being stationary. Finally, $U = T$ a.s.. The relation $U < T$ indeed implies that for sufficiently large m $d(U,T) > 1/m$; hence for all sufficiently large p $d(U^p,T) > 1/m$ and hence for every sufficiently large p there exists $n \geq p$ such that $d(T_n^{n'},T) > 1/n$. Since the sequence (T_n) is stationary, this is equivalent for large n to $d(T_n^{n'},T_n) > 1/n$ and we see that

$$\{U < T\} \subset \limsup_n \{d(T_n^{n'},T_n) > 1/n\} \quad \text{a.s.}$$

which is a negligible set By the Borel-Cantelli lemma. All this means that the sequence (U_p) foretells T a.s.

(d) Let (S^n) and (T^n) be two sequences foretelling S and T a.s. We set

$$U_n^m = n \wedge S^n_{\{S^n < T^m\}}$$

For fixed m we thus construct an increasing sequence of (wide sense) stopping times whose limit U^m is equal to $+\infty$ on $\{T^m = 0\}$ and is a.s. equal to $S_{\{S \leq T^m\}}$ on $\{T^m > 0\}$. On the other hand, the sequence (U_n^m) foretells U^m a.s. and U^m belongs to $\mathfrak{U}$. But the sequence (U^m) is decreasing and stationary, so that its limit U belongs to $\mathfrak{U}$ by (c). We therefore denote by (V_n) a sequence which foretells U a.s. ; it is then immediate that $U = S_{\{S < T\}}$ a.s., and it only remains to set $V_n' = V_n \wedge S_{\{S \wedge T\}}$ to get a sequence foretelling $S_{\{S \wedge T\}}$ a.s.

(e) If the sequence (S_n) foretells S a.s., the sequence $(S_n \wedge T)$ foretells T a.s.

The following theorem is the crucial step leading to theorem 77. The arguments used are exactly those which will lead us later to the proof of the theorems on optional and predictable cross-sections (84 and 86). But here we do not need the theory of analytic sets and capacities to construct an auxiliary measure, a construction which will be more delicate in the case of cross-sections.

The statement is not definitive : it will be somewhat improved in no. 77.

76 THEOREM. Every predictable stopping time T belongs to $\mathfrak{U}$.

<u>Proof</u> : (a) Let $\mathcal{I}_0$ be the paving on $\mathbb{R}_+ \times \Omega$ consisting of all stochastic intervals $[\![S,T[\![$, where S and T are elements of $\mathcal{U}$ such that $S \leq T$, and let $\mathcal{I}$ be the closure of $\mathcal{U}$ under ($\cup$f). It is immediate that the complement of any element of $\mathcal{I}_0$ is the union of two elements of $\mathcal{I}_0$($[\![S,T[\![^c = [\![0,S[\![\cup [\![T,\infty[\![$), and that the intersection of two elements of $\mathcal{I}_0$ belongs to $\mathcal{I}_0$ ($[\![S,T[\![\cap [\![U,V[\![= [\![S \vee U, S \vee U \vee (T \wedge V)[\![$), so that $\mathcal{I}$ is a Boolean algebra. It follows from remark 74 that the σ-field $\sigma(\mathcal{I})$ - which is also the monotone class generated by $\mathcal{I}$ - contains the predictable σ-field.

(b) Using only Lemma 75 (a), (b), (c) and (d), we show that <u>the debut of an element</u> B <u>of</u> $\mathcal{I}_\delta$ <u>is equal a.s. to an element of</u> $\mathcal{U}$.

We begin by showing that the debut of an element C of $\mathcal{I}$ belongs to $\mathcal{U}$. By definition of $\mathcal{I}$, we can write $C = C_1 \cup \ldots \cup C_i$, where the C_k belong to $\mathcal{I}_0$; then $D_C = D_{C_1} \wedge \cdots \wedge D_{C_i}$ and it suffices (75(a)) to show that the debut of an interval $C_k = [\![S,T[\![\in \mathcal{I}_0$ belongs to $\mathcal{U}$ - but this debut is equal to $S_{\{S < T\}}$ and this then follows from 75 (d).

Next we represent B as the intersection of a decreasing sequence (B_n) of elements of $\mathcal{I}$. Let $\mathcal{D}$ be the set of elements of $\mathcal{U}$ smaller than D_B ; $\mathcal{D}$ is non-empty ($0 \in \mathcal{D}$), $\mathcal{D}$ is closed under ($\vee$c), hence there exists an increasing sequence (H_n) of elements of $\mathcal{D}$ whose upper envelope $H \in \mathcal{U}$ is a.s. equal to the essential upper envelope of $\mathcal{D}$. Let $C_n = B_n \cap [\![H,\infty[\![$ and let S_n be the debut of C_n. C_n belongs to $\mathcal{I}$ and hence S_n belongs to $\mathcal{U}$. Since $H \leq D_B$, we have $B_n \supset C_n \supset B$ and hence $\bigcap_n C_n = B$. Then $S_n \geq H$, $S_n \leq D_B$ and hence the fact that H is the essential upper envelope of $\mathcal{D}$ implies that $S_n = H$ a.s. Since the graph of S_n is contained in C_n, the graph of H is contained a.s. in $\bigcap_n C_n = B$, so $H \geq D_B$ a.s. and finally $H = D_B$ a.s.

(c) We consider the following measure on $\mathbb{R}_+ \times \Omega$

$$\mu(A) = \int I_A(T(\omega),\omega) I_{\{T < \infty\}}(\omega) \mathbb{P}(d\omega) \quad A \in \mathcal{B}(\mathbb{R}_+) \times \mathcal{F}^0.$$

This is a bounded measure of total mass $\mathbb{P}\{T < \infty\}$, carried by the graph $[\![T]\!]$. This graph is predictable and hence belongs to $\sigma(\mathcal{I})$. By a classical theorem of measure theory (1), for all $\varepsilon > 0$ there exists an element B_ε of I_δ contained in $[\![T]\!]$ and such that $\mu([\![T]\!] \setminus B_\varepsilon) \leq \varepsilon$. Let T^ε be an element of $\mathcal{U}$ equal a.s. to the debut of B ((b) above) ; the graph of T^ε is a.s. contained in that of T, hence the stopping time $T^\varepsilon_{\{T=T^\varepsilon\}}$ is a.s. equal to T, it belongs to $\mathcal{U}$ by 75 (e) and its graph is entirely contained in that of T. We henceforth denote this stopping time by T^ε ; we have

$$T^\varepsilon \in \mathcal{U},\ T^\varepsilon = T \text{ on } \{T^\varepsilon < \infty\},\ \mathbb{P}\{T^\varepsilon < \infty\} \geq \mathbb{P}\{T^\varepsilon < \varepsilon\} - \varepsilon.$$

---

(1) For us, with Choquet's theorem at our disposal, this is also a result on capacitability concerning the paving $\mathcal{I}$ and the capacity μ^*, where every element of $\mathcal{P} \subset \sigma(\mathcal{I})$ is $\mathcal{I}$-analytic.

Let $S_n = T^{1/2} \wedge \ldots \wedge T^{1/n}$; then $S_n \in \mathcal{U}$, the sequence (S_n) is decreasing and stationary and its limit S is a.s. equal to T. We then conclude using 75 (c) and (e).

We now reach the definitive result on sequences foeretelling a predictable stopping time. The refinement concerning approximation by elementary predictable times is due to Chung [2]. For greater clarity, the statement repeats the definition of sequences which foretell T a.s.

(77) THEOREM. (a) Let T be a predictable stopping time of the family $(\mathcal{F}_t^o)$. There exists an increasing sequence (T_n) of predictable stopping times which are bounded above by T on the whole of Ω, such that $\lim_n T_n = T$ $\mathbb{P}$-a.s. and $T_n < T$ $\mathbb{P}$-a.s. for all n on $\{T > 0\}$. It can futher be assumed that each T_n takes its values a.s. in the set of dyadic numbers.

(b) If the usual conditions are satisfied, the "a.s." can be omitted in the above statement.

Proof : We shall suppose that T is everywhere >0 ; this is not a restriction since the set $\{T > 0\} = \Omega'$ belongs to $\mathcal{F}_{0-}^o$ and we may argue on Ω' with the induced law and then set $T_n = 0$ on $\{T = 0\}$.

(a) Let (S_n) be an increasing sequence of wide sense stopping times bounded above by T, which foretells T a.s. (76) and for all p let S_n^p be the dyadic approximation of S_n by higher values :

$$S_n^p = (k+1)2^{-p} \Leftrightarrow S_n \in [k2^{-p}, (k+1)2^{-p}[,\ S_n^p = \infty \Leftrightarrow S_n = \infty .$$

We have noted in no. 57 that S_n^p is a stopping time and in no. 72 that it is even predictable. As $S_n < T$ a.s., $\lim_n S_n^p = S_n$ and for all n there exists an integer $n' > n$ such that

(77.1) $$\mathbb{P}\{T \le S_n^{n'}\} < 2^{-n}.$$

Then we set $U_n = \inf_{m \ge n} S_m^{m'}$. Since $S_m^{m'} \le T + 2^{-m'}$, U_n is everywhere bounded above by T. The sequence (U_n) is increasing and the inequality $S_m^{m'} \ge S_m$ implies $U_n \ge S_n$, hence $\lim_n U_n \ge T$ a.s. and finally $\lim_n U_n = T$ a.s. Finally the Borel-Cantelli Lemma and (77.1) imply that, except on a negligible set, $S_m^{m'} < T$ for all sufficiently large m and hence $U_n < T$ for all n. Thus the sequence (U_n) is a sequence of stopping times in the wide sense which foretells T a.s.

But we can do better : let N be the negligible set outside which $\lim_n U_n = T$, $U_n < T$ for all n and let $\omega \in N^c$. The subset of $\bar{\mathbb{R}}$ consisting of the points $S_m^{m'}(\omega)$ $(m \ge n)$ and $T(\omega)$ is compact, since the sequence $(S_p^{p'})$ lying between U_p and $T + 2^{-p'}$ converges to T. Hence it contains its least upper bound $U_n(\omega)$. Since this is not equal to $T(\omega)$ by hypothesis, it is equal to one of the $S_m^{m'}(\omega)$. It follows immediately that $U_n(\omega)$ is a dyadic number.

We now construct a predictable stopping time $V_n \le T$ which is equal a.s. to U_n. First we note that the decreasing sequence of predictable stopping times

$$U^k = \lim_{n+k\ge m\ge n} S_m^{m}$$

whose limit is U_n, is a.s. stationary by the above. We write $V^k = U^k_{A_k} \wedge T$, where $A_k = \{U^k = U^{k+i}$ for all $i\}$; the V^k are decreasing, are bounded above by T, form an everywhere stationary sequence and are predictable by 73 (b). Their limit V_n is bounded above by T, is equal a.s. to U_n and is predictable by 73 (d).

If we now write $T_n = V_1 \vee V_2 \vee \ldots \vee V_n$, we have constructed an increasing sequence of predictable stopping times which are bounded above by T and are equal a.s. to the U_n - and hence a.s. have dyadic values and foretell T a.s.

(b) Under the usual conditions, all the negligible sets belong to $\mathcal{F}_{0-}$ (assuming this would simplify a lot the above proof, since then a stopping time which is a.s. equal to a predictable stopping time is itself predictable). Let N be the negligible subset of $\{T > 0\}$ where the T_n do not converge to T or are not $<T$ or do not have dyadic values. We modify T_n on N by giving it the value

$$T_n(\omega) = k2^{-n} \text{ if } k2^{-n} < T(\omega) \le (k+1)2^{-n},\ T_n(\omega) = 2^n \text{ if } T(\omega) = +\infty.$$

Since $\mathcal{F}_{0-}$ contains all the negligible sets, T_n still is a predictable time after this modification and the theorem is established.

> The reader can now see why the definition we adopted for predictable stopping times (69) is more satisfactory than the one which uses foretelling sequences : the notion of a predictable stopping time of the family $(\mathcal{F}^0_t)$ does not depend on the choice of a law $\mathbb{P}$ on Ω ; if T is predictable, then for every law $\mathbb{P}$ there exists a sequence which foretells T a.s., but this sequence depends on the law $\mathbb{P}$. (We could also have taken as definition the existence of a sequence of stopping times of $\mathcal{F}^0_t$ which foretell T <u>everywhere</u> on $\{T > 0\}$, a property which is also independent of $\mathbb{P}$, but we would have been bothered in n°73 with a number of "a.s"!)
>
> The following statement should be compared to no. 59.

THEOREM. <u>Let</u> $(\mathcal{F}_t)$ <u>be the usual augmentation of</u> $(\mathcal{F}^0_t)$ <u>and let</u> T <u>be a predictable time of</u> $(\mathcal{F}_t)$. <u>Then there exists a predictable time</u> T' <u>of</u> $(\mathcal{F}^0_t)$ <u>which is equal to</u> T <u>a.s.</u> <u>Further, for all</u> $A \in \mathcal{F}_{T-}$ <u>there exists</u> $A' \in \mathcal{F}^0_{T-}$ <u>such that</u> $A = A'$ <u>a.s.</u> 78

<u>Proof</u> : The set $\{T = 0\}$ belongs to $\mathcal{F}_{0-}$; we choose a set $H \in \mathcal{F}^0_{0-}$ which differs from $\{T = 0\}$ by a negligible set and modify T on H^c - without changing the notation - by replacing it by $+\infty$ on $H^c \cap \{T = 0\}$. Working henceforth in H^c instead of Ω, we can reduce the problem to the case where T is everywhere >0. We write $T' = 0$ on H and forget about H.

Let (T^n) be a sequence of stopping times of $(\mathcal{F}_t)$ which foretells T. For all n let (R^n) be a stopping time of the family $(\mathcal{F}^0_{t+})$ such that $T^n = R^n$ a.s. (59). Replacing R^n by $R^1 \vee \ldots \vee R^n$ if necessary, we can assume that the sequence (R^n) is increasing and denote its limit by R. Let $A_n = \{R^n = 0\} \cup \{R^n < R\}$. The A_n decrease, hence the stopping times of $(\mathcal{F}^o_{t+})$ $S_n = R^n_{A_n} \wedge n$ increase and the sequence S_n foretells everywhere its limit T', which is strictly positive. By 71 T' is a predictable stopping time of the family $(\mathcal{F}^o_t)$ and $T' = T$ a.s.

Then an argument using monotone classes based on 74 shows the following : for every predictable process $(X_t)_{t \in [0,\infty]}$ of the family $(\mathcal{F}_t)$ there exists a process $(X'_t)_{t \in [0,\infty]}$ which is indistinguishable from (X_t) and is predictable with respect to $(\mathcal{F}^o_t)$. We leave the details to the reader.

Finally let $A \in \mathcal{F}_{T-}$; we choose (X_t) to be predictable such that $X_T = I_A$ (67(b)) and then (X'_t) indistinguishable of (X_t) as above ; the required event A' is $\{X'_{T'} = 1\}$.

Classification of stopping times

Up to no. 83 (included), the filtration $(\mathcal{F}_t)$ satisfies the usual conditions.

79 We introduce some notation. Given a stopping time T, $\mathcal{S}(T)$ denotes the set of increasing sequences (S_n) of stopping times bounded by T. For every sequence $(S_n) \in \mathcal{S}$, we write

(79.1) $A[(S_n)] = \{\lim_n S_n = T,\ S_n < T \text{ for all } n\} \cup \{T = 0\}$.

If we forget about $\{T = 0\}$ (recall that 0 plays the devil in this theory), $A[(S_n)]$ is the set on which the sequence (S_n) foretells T. We note that $A[(S_n)]$ belongs to $\mathcal{F}_{T-}$: let $S = \lim_n S_n$; $A[(S_n)]$ is the union of $\{T = 0\}$, which is $\mathcal{F}_{T-}$-measurable by 56 (a), and of the set $(\bigcap_n \{S_n < T\}) \setminus \{S < T\}$, which is $\mathcal{F}_{T-}$-measurable by 56 (c). We denote by A(T) a representative of the essential union of the sets $A[(S_n)]$ and by I(T) the complement of A(T). Note that A(T) always contains $\{T = 0\}$ and $\{T = +\infty\}$ (a.s).

80 DEFINITION. T is said to be accessible if $A(T) = \Omega$ a.s. and totally inaccessible if $A(T) = \{T = +\infty\}$ a.s.

In other words, T is totally inaccessible if $T > 0$ a.s. and if for every sequence (T_n) of stopping times that increases to T the event $\{\lim_n T_n = T, T < \infty\}$ has zero probability. This means that one cannot localize T exactly using a sequence of "precursory signs". Hence "totally inaccessible" is just the opposite of "predictable". However, there is an important difference - whereas the notion of a predictable stopping time can be defined without a probability law, that of a totally inaccessible stopping time is relative to a given law $\mathbb{P}$.

It is also necessary to understand the difference between accessible and predictable : if T is accessible, Ω is a.s. the union of sets A_k, on each of which T is foretold by some sequence (S^k_n). But outside A_k the sequence (S^k_n) may behave badly and it is impossible - if T is not predictable - to rearrange all these sequences into a single one which foretells T a.s. on the whole of Ω .

We shall see in §4 examples of totally inaccessible stopping times and of non-predictable accessible times.

The following theorem contains the essential results on the classification of stopping times. For more detail and in particular for the study of accessible times, see Dellacherie [1].

THEOREM. (a) T is accessible, if and only if $[\![T]\!]$ is a.s. contained in a countable union of graphs of predictable stopping times. (81)

(b) T is totally inaccessible, if and only if $\mathbb{P}\{S = T < \infty\} = 0$ for every predictable stopping time S.

(c) The stopping time $T_{A(T)}$ is accessible and the stopping time $T_{I(T)}$ is totally inaccessible. This decomposition is unique in the following sense : if U and V are two stopping times with U accessible, V totally inaccessible, $U \wedge V = T$ and $U \vee V = +\infty$, then $U = T_{A(T)}$ and $V = T_{I(T)}$ a.s.

Proof : (a) Suppose that T is accessible. There then exist sequences $(S_n^k)_{n \in \mathbb{N}}$ belonging to $\mathcal{S}(T)$ such that Ω is a.s. the union of the sets $A[(S_n^k)]$. For every k we set $S^k = \lim_n S_n^k$ and then

$$R_n^k = (S_n^k)_{\{S^k < S^k\}} \wedge n.$$

We thus define an increasing sequence of stopping times which foretells its limit R^k everywhere on Ω, and $[\![T]\!]$ is a.s. contained in the union of the graphs of the predictable stopping times R^k and 0.

Conversely, suppose that $[\![T]\!]$ is a.s. contained in the union of the graphs of a sequence of predictable stopping times T^k. For each k let (T_n^k) be a sequence which foretells T^k and let $S_n^k = T_n^k \wedge T$; then the sequence (S_n^k) belongs to $\mathcal{S}(T)$, the set $\{T = T^k < \infty\}$ is contained in $A[(S_n^k)]$ and hence T is accessible.

(b) Let S be a predictable time, (S_n) be a sequence foretelling S and let $S_n' = S_n \wedge T$; the sequence (S_n') belongs to $\mathcal{S}(T)$ and $\{S = T\} \subset A[(S_n')]$; since T is totally inaccessible, $A[(S_n')]$ is a.s. contained in $\{T = \infty\}$ and $\mathbb{P}\{S = T < \infty\} = 0$.

Conversely, suppose that $\mathbb{P}\{R = T < \infty\} = 0$ for every predictable time R. Then in particular $\mathbb{P}[T = 0] = 0$. Let (S_n) be a sequence belonging to $\mathcal{S}(T)$. As in part (a) we set $S = \lim_n S_n$, $R_n = (S_n)_{\{S_n < S\}} \wedge n$ and $R = \lim_n R_n$; R is predictable, hence $\mathbb{P}\{R = T < \infty\} = 0$ and this means that $A[(S_n)] \subset \{T = \infty\}$ a.s. ; hence T is totally inaccessible.

(c) The argument in (a) shows that $[\![T_{A(T)}]\!]$ is contained in a union of graphs of predictable times and hence $T_{A(T)}$ is accessible. The first part of the argument in (b) shows that, if S is a predictable time, the set $\{S = T < \infty\}$ is a.s. contained in $A(T)$, so that $\mathbb{P}\{T_{I(T)} = S < \infty\} = 0$ and $T_{I(T)}$ is totally inaccessible by (b). We leave uniqueness to the reader.

Quasi-left-continuous filtrations

We have mentioned above that the left-continuity condition $\mathcal{F}_t = \mathcal{F}_{t-}$ for all t isn't sufficient to imply interesting results. If the same condition holds not only at constant times, but at all predictable times, a situation which happens quite frequently, the classification of stopping times is much simplified. We prove only one result here : for more details, see Dellacherie [1].

82 DEFINITION. Let $(\mathcal{F}_t)$ be a filtration satisfying the usual conditions. $(\mathcal{F}_t)$ is said to be quasi-left-continuous if $\mathcal{F}_T = \mathcal{F}_{T-}$ for all predictable times T (and in particular for T = 0 and T = t).

The quasi in the phrase "quasi-left-continuous" comes from the "quasi-left-continuity" property of some Markov processes (with right-continuous paths) or of some martingales. It does not imply the existence of a stronger notion of a "left-continuous" family of σ-fields. In particular, equality for all t of the σ-fields $\mathcal{F}_t$ and $\mathcal{F}_{t-}$ is a weaker property than quasi-left-continuity and must not be called left-continuity.

Moreover, there exist quasi-left-continuous families such that the equality $\mathcal{F}_{T-} = \mathcal{F}_T$ does not hold at some (non-predictable) stopping times T.

83 THEOREM. Suppose that $(\mathcal{F}_t)$ is quasi-left-continuous. Then

(a) every accessible stopping time is predictable,

(b) for every increasing sequence (T_n) of stopping times, with $\lim_n T_n = T$, we have

$$\mathcal{F}_T = \bigvee_n \mathcal{F}_{T_n} .$$

Proof : (a) Let T be an accessible stopping time ; there exists a sequence of graphs of predictable stopping times R_n whose union contains T. For all n, the set $\{R_n = T\}$ belongs to $\mathcal{F}_{R_n}$, which is equal to $\mathcal{F}_{R_n -}$; we then write $S_n = (R_n)_{\{R_n = T\}}$: S_n is predictable by 73 (c). Then T is the limit of the stationary decreasing sequence $T_n = S_1 \wedge \ldots \wedge S_n$ of predictable times ; hence it is predictable (73 (d)).

(b) Let $H = \{T_n < T \text{ for all } n\} \in \mathcal{F}_T$ and let $S = T_H$; S is predictable (it is foretold by the sequence $n \wedge S_n$, where $S_n = (T_n)_{\{T_n < T\}}$). Let $A \in \mathcal{F}_T$; we show separately that $A \cap H$ and $A \cap H^c$ belong to $\bigvee_n \mathcal{F}_{T_n}$.

1. $A \cap H^c$ is the union of the sets $A \cap \{T \le T_n\} \in \mathcal{F}_{T_n}$ (56.1).

2. $A \cap H$ belongs to $\mathcal{F}_T$ and hence to $\mathcal{F}_S$ since $T \le S$. As $\mathcal{F}_S = \mathcal{F}_{S-}$ by hypothesis, $A \cap H = A \cap H \cap \{S \le T\}$ belongs to $\mathcal{F}_{T-}$ by 73 (c) and hence to $\bigvee_n \mathcal{F}_{T_n -}$ by 56 (d). Then it belongs a fortiori to $\bigvee_n \mathcal{F}_{T_n}$.

Conversely, if every accessible time is predictable, the filtration is quasi-left-continuous. For if T is predictable and A belongs to $\mathcal{F}_T$, T_A is accessible (81) and hence predictable and then $A \in \mathcal{F}_{T-}$ (73).

The cross-section theorem [1]

We are coming to the most important theorems in this paragraph : the optional cross-section theorem and the predictable cross-section theorem. The proof which we give (taken from Dellacherie [1]), derives them in a simple way from the "ordinary" (i.e., non filtered) cross-section theorem, itself an easy application of Choquet's theorem. So we hope to dissipate the fear created by the obscure proofs of the earlier authors (cf. the first edition of this book). If we succeed in convincing probabilists that everything here is trivial, we'll be perfectly happy.

THEOREM. Let $(\Omega,\mathcal{F},\mathbb{P})$ be a complete probability space with an arbitrary filtration $(\mathcal{F}_t^0)$ 84
and let A be an optional set. For every $\varepsilon > 0$ there exists a stopping time T of the family $(\mathcal{F}_t^0)$ with the following properties :

(a) for all ω such that $T(\omega) < \infty$, $(T(\omega),\omega) \in A$;

(b) $\mathbb{P}\{T < \infty\} \geq \mathbb{P}(\pi(A)) - \varepsilon$ [2], where $\pi(A)$ is the projection of A onto Ω.

Proof : (a) We begin by choosing, using III.44, a measurable cross-section of A, i.e. an $\mathcal{F}^0$-measurable real-valued r.v. R (in general not a stopping time) such that

$$R(\omega) < \infty \Rightarrow (R(\omega),\omega) \in A$$
$$\mathbb{P}\{R < \infty\} = \mathbb{P}(\pi(A))$$

and we use it to construct a measure μ on $\mathbb{R}_+ \times \Omega$: if G is an element of $\mathcal{B}(\mathbb{R}_+) \times \mathcal{F}^0$

$$\mu(G) = \int I_G(R(\omega),\omega) I_{\{R < \infty\}}(\omega) \mathbb{P}(d\omega).$$

This measure is carried by A and its mass is equal to $\mathbb{P}(\pi(A))$. This is the only place where we use a result from Chapter III. The rest of the proof will be elementary measure theory.

(b) We now copy the proof of no. 76, with $\mathcal{U}$ this time denoting the set of all stopping times of $\mathcal{F}_t^0$, $\mathcal{J}_0$ the paving consisting of the intervals $[S,T[$, with $S \leq T$ and $S, T \in \mathcal{U}$, and $\mathcal{J}$ the closure of $\mathcal{J}_0$ under $(\cup f)$. As in no. 76, $\mathcal{J}$ is a Boolean algebra which generates the optional σ-field (64) and the debut of an element B of $\mathcal{J}_\delta$ is a.s. equal to an element of $\mathcal{U}$. Again as in no. 76, by a classical theorem of measure theory, there exists a set $B \in \mathcal{J}_\delta$ contained in A such that $\mu(B) \geq \mu(A) - \varepsilon$. We denote by S an element of $\mathcal{U}$ which is a.s. equal to the debut of B. Because of the "a.s." the graph of S does not pass through B everywhere, so that we take $T = S_L$, where $L = \{\omega : (S(\omega),\omega) \in B\}$, which belongs to $\mathcal{F}_S^0$ by 64. T is then the required stopping time.

The same argument now gives us the predictable cross-section theorem.

---

(1) The completeness enables us to write $\mathbb{P}$ here instead of $\mathbb{P}^*$. This is its only use.

(2) English speaking authors often say "section theorems" according to the French terminology. The standard (topological) english terminology is rather "selection theorems" or "uniformization theorems". We have adopted a middle course.

85 THEOREM. Let $(\Omega,\mathcal{F},\mathbb{P})$ be a complete probability space with an arbitrary filtration $(\mathcal{F}_t^o)$ and let A be a predictable set. For every number $\varepsilon > 0$ there exists a predictable stopping time T of the family $(\mathcal{F}_t^o)$ with the following properties

(a) for all ω such that $T(\omega) < \infty$, $(T(\omega),\omega) \in A$;

(b) $\mathbb{P}\{T < \infty\} \geq \mathbb{P}(\pi(A)) - \varepsilon$, where $\pi(A)$ is the projection of A onto Ω.

Proof : We construct a measure μ carried by A of mass $\mathbb{P}(\pi(A))$, as in the first part of the preceding proof. Then we again copy the proof of 76 : $\mathcal{U}$ now denotes the set of all predictable stopping times, $\mathcal{J}_0$ is the set of intervals $[\![S,T[\![$ $(S,T \in \mathcal{U})$ and $\mathcal{J}$ is the closure of $\mathcal{J}_0$ under $(\cup f)$, a Boolean algebra which generates the predictable σ-field (73). As in no. 76, it can be verified that the debut of an element of $\mathcal{J}_\delta$ is a.s. equal to an element of $\mathcal{U}$: the proof of 76 is unchanged, as it depends only on the properties (a), (b), (c) and (d) of no. 75, which are satified by predictable stopping times (73). Again we choose $B \in \mathcal{J}_\delta$ contained in A and such that $\mu(B) \geq \mu(A) - \varepsilon$, we denote by S a predictable time a.s. equal to D_B and we get the required predictable time by taking $T = S_L$, where $L = \{\mu: (S(\omega),\omega) \in B\}$ belongs to $\mathcal{F}^o_{S-}$ (73).

The following is the most frequent application of the cross-section theorems.

86 THEOREM. Let (X_t) and (Y_t) be two optional (resp. predictable) processes. Suppose that for every stopping time (resp. predictable stopping time) T, $X_T = Y_T$ a.s. on $\{T < \infty\}$. Then the two processes are indistinguishable.

Proof : For simplicity we deal with the optional case. It is sufficient to show that for all $\varepsilon > 0$ the optional set $A = \{(t,\omega) : X_t(\omega) > Y_t(\omega) + \varepsilon\}$ is evanescent. But if it were not, it would admit a cross-section by a stopping time T such that $\mathbb{P}\{T < \infty\} > 0$, contradicting the hypothesis.

87 REMARKS. (a) This statement extends immediately to processes with values in a separable metrizable space E (consider the real processes $(f \circ X_t)$ and $(f \circ Y_t)$, where f runs through a countable set of Borel functions generating the σ-field $\mathcal{B}(E)$).

(b) Suppose that for every stopping time (resp. predictable stopping time) T the random variables $X_T I_{\{T < \infty\}}$ and $Y_T I_{\{T < \infty\}}$ are integrable and have the same expectation. Then the same proof shows that the two processes are indistinguishable.

(c) Let S be a positive r.v. Then S is a stopping time (resp. predictable stopping time) if (and only if) $[\![S]\!]$ is an optional (resp. predictable) set. Indeed, there then exists an optional (resp. predictable) time T_n which is a cross-section of $[\![S]\!]$, such that $\mathbb{P}\{S \neq T_n\} < 1/n$ and then $[\![S,\infty[\![$ is indistinguishable from $\bigcup_n [\![T_n,\infty[\![$ which is optional (resp. predictable).

(d) Let H be a predictable set. If the graph of the debut D_H is contained in H (in particular if H is right-closed), D_H is predictable. For then $[\![D_H]\!] = H \setminus]\!]D_H,\infty[\![$ which is a predictable set, and we apply the preceding remark.

88 We end with a result which contains nothing essentially new, but is an excellent exercise on the whole paragraph : it makes precise the structure of some measurable sets with countable sections, which often occur as "exceptional sets" when modifying

processes (cf. 66, no. 88B below, and 91 (a)).

THEOREM. Let $(\Omega,\mathcal{F},\mathbb{P})$ be a complete probability space with an arbitrary filtration $(\mathcal{F}_t^o)$.

(a) A progressive set contained in a countable union of graphs of optional times is the union of a sequence of disjoint graphs of optional times, and is therefore optional.

(b) A predictable set contained in a countable union of graphs of predictable (resp. optional) times is equal to (resp. indistinguishable from) the union of a sequence of disjoint graphs of predictable times.

(c) Let H be a subset of $\bar{\mathbb{R}}_+ \times \Omega$ contained in a countable union of graphs of positive random variables. Then H has an essentially unique (i.e. unique up to indistinguishability) decomposition

$$H = H' \cup H''$$

where H' and H" are disjoint measurable sets, H' is contained in a countable union of graphs of optional times and H" intersects every graph of an optional time along an evanescent set. Further, if H is optional (resp. predictable), then H is indistinguishable from the union of a sequence of (disjoint) graphs of optional (resp. predictable) times.

Proof : Let L be a subset of $\mathbb{R}_+ \times \Omega$ contained in the union of the graphs of a sequence (S_n) of positive r.v. We set $T_1 = S_1$ and, for $n \geq 2$, $T_n = S_n$ on $\{S_1 \neq S_n, \ldots, S_{n-1} \neq S_n\}$ and $T_n = +\infty$ otherwise : the T_n are positive r.v. with disjoint graphs and L is contained in the union of these graphs. If the S_n are optional (resp. predictable) times, the T_n are also optional (resp. predictable) times by 53 and 56 (c) (resp. 73 (b) and (c)) ; if further L is progressive (resp. predictable), then, for each n, $[\![T_n]\!] \cap L$ is the graph of an optional (resp. predictable) time. The progressive case has been shown earlier (no. 66) ; the predictable case can be treated similarly : the indicator (X_t) of L then is indeed a predictable process and $A_n = \{X_{T_n} = 1, T_n < \infty\}$ belongs to $\mathcal{F}_{T-}$ by 67 (b), so $[\![T_n]\!] \cap L$ is the graph of the predictable time equal to T_n on A_n and to $+\infty$ on A_n^c . We have thus established (a) and part of (b). The other part of (b) follows from (c) which we now establish.

Let $\mathcal{U}$ be the set of optional (resp. predictable) times and for every positive r.v. Z let V(Z) denote a representative of the essential union of the sets $\{Z = T\}$, where T runs through $\mathcal{U}$. We define two positive r.v. Z' and Z" by $Z' = Z$ on $V(Z)$, $Z' = +\infty$ on $V(Z)^c$ and $Z'' = Z$ on $V(Z)^c$, $Z'' = +\infty$ on $V(Z)$. Paraphrasing 81, we could call Z': the $\mathcal{U}$-accessible part of Z and Z" the $\mathcal{U}$-totally inccessible part of Z : the graph of Z' is contained in a countable union of graphs of elements of $\mathcal{U}$ and $\mathbb{P}\{Z'' = T < \infty\} = 0$ for all $T \in \mathcal{U}$.

Now let H be a subset of $\mathbb{R}_+ \times \Omega$ contained in the union of the graphs of a sequence (Z_n) of positive r.v. By the above, with $\mathcal{U}$ the set of optional times, there exists a sequence (T_n) of optional times such that $\bigcup_n [\![Z'_n]\!]$ is contained in $\bigcup_n [\![T_n]\!]$, and we have seen earlier that the graphs $[\![T_n]\!]$ can be assumed to be disjoint. We then write

$$H' = H \cap_n (\bigcup [\![Z'_n]\!]) = H \cap_n (\bigcup [\![T_n]\!]),\ H'' = H \cap_n (\bigcup [\![Z''_n]\!]) = H \setminus (\bigcup_n [\![T_n]\!]).$$

The sets H' and H" constitute a decomposition of H with the properties required by (c) and clearly this decomposition is unique up to indistinguishability. If H is optional, so are H' and H" ; then H" is evanescent by the cross-section theorem 84 and H is indistinguishable from H', which is a countable union of disjoint graphs of optional times by (a). If H is predictable, we return to the decomposition of H with $\mathcal{U}$ this time the set of predictable times. We get another decomposition of H where H' is predictable and is contained in a countable union of graphs of predictable times, and H" is predictable and intersects every graph of a predictable time along an evanescent set : we then conclude the proof using the cross-section theorem 85 and the first part of (b).

We shall show in the appendix to Chapter IV that a measurable subset H of $\mathbb{R}_+ \times \Omega$, such that the section $H(\omega)$ is at most countable for all $\omega \in \Omega$, is contained in a countable union of graphs of positive random variables.

REMARKS. (a) If the filtration $(\mathcal{F}^o_t)$ is complete, every function which is positive on Ω and zero $\mathbb{P}$-a.s. is a predictable time. It is easily deduced that in (c), if the set H is optional (resp. predictable), then it is the union of a sequence of disjoint graphs of optional (resp. predictable) times.

(b) The reader will have noted that the above proof says more than the statement : in assertion (c), H also has a decomposition relative to predictable times.

(c) We shall see in §4 (no. 91 (b)) that there exists a progressive set contained in a countable union of graphs of (non-optional) random variables, which is not optional and even contains no graph of a stopping time.

The following numbers are complements to 88, which were found to be useful when writing some parts of chapter VI. Hence their numbering as 88B, C and D. The first and last one concern explicit representations of some sets with countable sections as unions of graphs of stopping times, and the intermediate one is a consequence of 88B, a very convenient criterion for predictability of an optional process.

88B THEOREM. <u>Let</u> (X_t) <u>be a real valued, adapted r.c.l.l. process</u>. <u>We make the convention that</u> $X_{0-} = X_0$ <u>and set</u>

$$U = \{(t,\omega) : X_t(\omega) \neq X_{t-}(\omega)\}$$

Then U is a countable union of disjoint graphs of stopping times, which may be chosen predictable if (X_t) is predictable.

Proof : Everything can of course be deduced from theorem 117 of the appendix, but it is better to give an elementary proof.

Let us set $U_n = \{(t,\omega) : |X_t(\omega) - X_{t-}(\omega)| > 2^{-n}\}$ $(n \geq 0)$, then $V_0 = U_0$, $V_n = U_n \setminus U_{n-1}$ $(n > 0)$; the sets V_n are optional (predicatable if so is (X_t)) and disjoint. Next we set

$$D_n^1(\omega) = \inf\{t : (t,\omega) \in V_n\} \quad , \quad D_n^{k+1}(\omega) = \inf\{t > D_n^k(\omega) : (t,\omega) \in V_n\}$$

so that D_n^i is the i-th jump of (X_t) whose size lies between 2^{-n} and 2^{-n+1}. Since (X_t) has r.c.l.l. paths, V_n has no finite cluster point, and the stopping times D_n^i enumerate all points of V_n. If (X_t) is predictable, then the D_n^i are predictable according to 87d). Finally, it only remains to reorder the double sequence D_n^i into an ordinary sequence (T_n).

REMARK. The conclusion still holds for a process taking values in a metrizable separable space E : one just imbeds E into $[0,1]^{\mathbb{N}}$ and applies the statement to each coordinate process, then the procedure at the beginning of the proof of 88 to turn the stopping times into disjoint ones.

The same remark (with the same argument) applies to the following application of 88B, which interrupts our discussion of sets with countable sections.

THEOREM. Let $X = (X_t)$ be a real-valued adapted r.c.l.l. process. Then X is predictable if and only if the following two conditions are satisfied. 88C

1) For every totally inaccessible stopping time T, X_T and X_{T-} are a.s. equal on $\{T < \infty\}$.

2) For every predictable time T, X_T is $\mathcal{F}_{T-}$-measurable on $\{T < \infty\}$.

Proof : Assume X is predictable. Then 2) is satisfied for every (optional) T according to 67. On the other hand, from 88B the set $U = \{(t,\omega) : X_t(\omega) \neq X_{t-}(\omega)\}$ is the union of countably many graphs of predictable times, whence 1) follows at once.

Conversely, assume 1) and 2) are satisfied. We represent U as a countable union of graphs of optional times S_n, then decompose S_n into its totally inaccessible part S_n^i and its accessible part S_n^a (81. (c)). According to 1) S_n^i is a.s. equal to $+\infty$, so that $S_n = S_n^a$ is accessible. Then the graph of S_n is contained in a countable union of graphs of predictable times S_{nk} $(k \in \mathbb{N})$ by virtue of 81 (a). Let V be the union of the graphs $[\![S_{nk}]\!]$. From 88 we may represent V as a union $[\![T_m]\!]$ of disjoint graphs of predictable times.

For every m, X_{T_m} and X_{T_m-} both are $\mathcal{F}_{T_m-}$-measurable : the first one according to 2), the second one from 67. So the same is true of $\Delta X_{T_m} = X_{T_m} - X_{T_m-}$. According to 67, there exists a predictable process (Y_t^m) such that $Y_{T_m}^m = \Delta X_{T_m}$ on $\{T_m < \infty\}$. On the other hand, the graph $[\![T_m]\!]$ is predictable. Denoting by X_- the process $(X_{t-})_{t \geq 0}$ $(X_{0-} = X_0)$, which is predictable by left-continuity, we have

$$X = X_- + \sum_m Y^m I_{[\![T_m]\!]}$$

and X is therefore predictable.

The following result looks like 88B, but for r.c. instead of r.c.l.l. processes, and it is less useful.

88D THEOREM. Let (X_t) be real-valued, adapted and right continuous. We make the convention that $X_{0-} = X_0$ and set

$$U = \{(t,\omega) : X_{t-} \text{ doesn't exist or } X_{t-}(\omega) \neq X_t(\omega)\} .$$

Then U is a countable union of disjoint graphs of stopping times, which may be chosen predictable if (X_t) is predictable.

Proof : To illustrate the possibilities offered by this chapter, we use a different method. We deal only with the predictable case.

We first show that U is predictable. To this end, we define

$$Y_t^+ = \limsup_{s \uparrow\uparrow t} X_s, \quad Y_t^- = \liminf_{s \uparrow\uparrow t} X_s$$

predictable according to 90. Then U is the union of the predictable sets $\{Y^+ \neq X\}$, $\{Y^- \neq X\}$.

To conclude we need only prove (according to 88) that U is contained - up to evanescent sets - into a countable union of graphs of positive random variables. Such r.v. (which turn out to be optional, but not necessarily predictable) are constructed as follows. We choose $\varepsilon > 0$ and set by transfinite induction

$$T_0^\varepsilon = 0, \quad T_{\alpha+1}^\varepsilon = \inf\{t > T_\alpha^\varepsilon : |X_t - X_{T_\alpha^\varepsilon}| > \varepsilon\}$$

$$T_\beta^\varepsilon = \sup_{\alpha < \beta} T_\alpha^\varepsilon \quad \text{if } \beta \text{ is a limit ordinal}$$

Since X is right continuous, we have $T_\alpha^\varepsilon < T_{\alpha+1}^\varepsilon$ on the set $\{T_\alpha^\varepsilon < \infty\}$. According to no. 8 in chapter 0, there exists a countable ordinal γ_ε such that $T_{\gamma_\varepsilon}^\varepsilon = +\infty$ a.s. Then U is contained in the union of all graphs $[\![T_\alpha^\varepsilon]\!]$ for $\varepsilon = 1/n$ $(n \in \mathbb{N})$ and $\alpha < \gamma_\varepsilon$: times t where the left hand limit doesn't exist appear among all T_β^ε corresponding to limit ordinals, if ε is small enough, while jump times occur among all $T_{\alpha+1}^\varepsilon$ for ε small enough.

REMARK. The result extends to processes with values in a metrizable separable space E, but the argument is slightly more delicate. We imbed E into $F = [0,1]^{\mathbb{N}}$, and remark that the result is trivial when X is considered as a F-valued process, but that the left limit may exist in F without existing in E. Then we decompose U into

$$U = \{X_{t-} \text{ doesn't exist in } E\} \cup \{X_{t-} \text{ exists in } E \text{ and } X_{t-} \neq X_t\}$$
$$= \{X_{t-} \text{ doesn't exist in } F\} \cup \{X_{t-} \text{ exists in } F \text{ and } X_{t-} \notin E\}$$
$$\cup\{X_{t-} \text{ exists in } E \text{ and } X_{t-} \neq X_t\}$$
$$= \{X_{t-} \text{ doesn't exists in } F\} \cup\{X_{t-} \text{ exists in } F \text{ and } X_{t-} \neq X_t\}$$

since (X_t) is an E-valued process. So we are really reduced to the same problem about F, which was shown above to be trivial.

4. EXAMPLES AND SUPPLEMENTS

Examples of optional and predictable processes

We return to the results in §2 on measurability, meaning now to establish that processes are optional or predictable instead of just progressive. The essential result is the following one (two statements are given, one for sets and one for processes, but they are really the same theorem).

89 THEOREM. Let $(\Omega,\mathcal{F},P)$ be a complete probability space with a filtration $(\mathcal{F}_t)$ which satisfies the usual conditions (with the convention $\mathcal{F}_{0-} = \mathcal{F}_0$). Let L be a progressive random set. Then

(a) the set L_1 of left accumulation points of L is predictable on $]0,\infty]$;

(b) the closure $\bar{L}$ of L is optional ;

(c) the set L_2 of right accumulation points of L is progressive on $[0,\infty[$.

90 THEOREM. With the same hypotheses, let $(X_t)_{t \in \mathbb{R}_+}$ be a real-valued progressive process. Then

(a) the process $U_t = \limsup_{s \uparrow\uparrow t} X_s$ is predictable on $]0,\infty]$;

(b) the process $V_t = \limsup_{s \to t} X_s$ is optional on $[0,\infty[$;

(c) the process $W_t = \limsup_{s \downarrow\downarrow t} X_s$ is progressive on $[0,\infty[$.

Proof : Assertions (c) of both statements are repetitions of (32-33). We first prove assertions (a) and (b) about sets. We set

$$A_t(\omega) = \sup\{s < t : (s,\omega) \in L\} \quad (t > 0 \text{ ; with the convention } \sup \emptyset = 0).$$

Since the σ-fields $(\mathcal{F}_t)$ are complete, and analyticity argument which we have often used [1] shows that A_t is $\mathcal{F}_t$-measurable. On the other hand, the paths $A_{\cdot}(\omega)$ are left-continuous and increasing. It first follows that the process $(A_t)_{t>0}$ is predictable. Then the right-hand limits (A_{t+}) exist everywhere and constitute a r.c.l.l. process adapted to the family $(\mathcal{F}_{t+}) = (\mathcal{F}_t)$ - and hence an optional process. We conclude the proof by noting that

$$L_1 = \{(t,\omega) : 0 < t < \infty, A_t(\omega) = t\}$$
$$\bar{L} = \{(t,\omega) : 0 < t < \omega, A_{t+}(\omega) = t\} \cup [\![D_L]\!].^{(2)}$$

We now pass easily from sets to processes : $U_t(\omega) \geq a$ if and only if for all $\varepsilon > 0$, t is a left accumulation point of the set $L_\varepsilon = \{(s,\omega) : X_s(\omega) > a - \varepsilon\}$. The argument for (V_t) is analogous.

(1) In no. 50 for example.

(2) D_L takes care of what happens for $t = 0$.

91 REMARKS. (a) Here are some supplements to the above statement, which are sometimes useful.

(1) If (X_t) is predictable, the process $U'_t = \limsup_{s\uparrow t} X_s$ is predictable ($U'_t = U_t \vee X_t$ for $t > 0$ and $U'_0 = X_0$). By way of symmetry we note that if (X_t) is progressive, the process $W'_t = \limsup_{s\downarrow t} X_s = W_t \vee X_t$ is progressive. This is of no interest, unlike the assertion about predictable processes.

(2) If (X_t) is optional, the process $V'_t = \limsup_{s\to t, s\neq t} X_s$ is optional. For $V'_t = U_t \vee W_t$ and $V'_t = \limsup_{s\to t} V'_s$. From the first equality it follows that (V'_t) is progressive and from the second that it is then optional. Note incidently that the set $\{V \neq V'\}$, which is optional, is contained in the union of the graphs of stopping times

$$D_r^{a,b} = \inf\{t \geq r : V'_t \leq a < b \leq V_t\}$$

where a,b run through all rationals and r through the positive rationals. This follows easily from the fact that V' is bounded above by V and that for fixed ω, a and b the set $\{t : V'_t(\omega) \leq a < b \leq V_t(\omega)\}$ is discrete. It then follows from 88 that $\{V \neq V'\}$ is the union of a sequence of disjoint graphs of stopping times.

(b) It is quite natural to ask whether statements (c) in the above theorems can be improved. For the initiated reader we show that this is impossible, even when (X_t) is a predictable process. Let (B_t) be a Brownian motion issuing from 0, with continuous paths, and $(\mathcal{F}_t)$ the filtration $(\sigma(B_s, s \leq t))$ suitable augmented. Let (X_t) be the indicator of the random set $M = \{(t,\omega) : B_t(\omega) = 0\}$; since (B_t) is continuous and hence predictable, the set M is closed and predictable (it can be shown that it is a.s. a perfect set without interior points). The process (W_t) then is the indicator of the set L of points of M which are not isolated from the right and the set $K = M \setminus L$ is the indicator of the set of points of M which are isolated from the right. K is therefore a progressive set which is discrete under the right topology and has countable sections [1]. Now let T be a stopping time whose graph passes through K ; its graph also passes through $M = \{t : B_t = 0\}$ and it follows from the strong Markov property of (B_t) that T is a.s. a right accumulation point of M on the set $\{T < \infty\}$. In other words, by the definition of K, every stopping time whose graph passes through K is a.s. infinite. Since K is not evanescent, it follows from Theorem 84 that K is not optional - and hence $L = M \setminus K$ is progressive and not optional and (W_t) is not optional although (X_t) is predictable.

There exists a progressive set with a.s. uncountable sections, and still containing no graph of a stopping time : for example the set

$$H = \{(t,\omega) : \limsup_{h\downarrow\downarrow 0} |B_{t+h}(\omega) - B_t(\omega)| \,/\, \sqrt{2\log\log \tfrac{1}{h}} < 1\}$$

(1) It is very easy to represent K explicitly as a union of a sequence of graphs.

see Knight [2], and another example in Dellacherie [7].

We now make theorems 17 and 38 more precise but without giving detailed proofs.

THEOREM. Let D be a countable dense subset of $\mathbb{R}_+$ and let $(X_t)_{t \in D}$ be a real-valued process defined on D such that X_t is $\mathcal{F}_t$-measurable for all $t \in D$. Then 92

(a) the process $U_t = \limsup_{s \uparrow\uparrow t, s \in D} X_s$ is predictable on $]0,\infty]$;

(b) the process $V_t = \limsup_{s \to t, s \in D} X_s$ is optional on $[0,\infty[$.

THEOREM. Let $(X_t)_{t \in \mathbb{R}_+}$ to a real progressive process. Then 93

(a) the process $U_t = \operatorname{ess} \limsup_{s \uparrow\uparrow t} X_s$ is predictable on $]0,\infty]$;

(b) the process $V_t = \operatorname{ess} \limsup_{s \to t} X_s$ is optional on $[0,\infty[$.

(These two statements reduce to 90 ; to prove 92, 90 is applied to the progressive process (Z_t) where $Z_t = X_t$ for $t \in D$ and $Z_t = -\infty$ for $t \notin D$; for 93, we know by 38 that the processes concerned are progressive, and we apply 90 to them. But in both cases assertions (a) can also be shown "with bare hands" even without the usual conditions on the family of σ-fields).

Stopping times on canonical spaces.

We now give - in a concrete case, without making any general theory - an interpretation of the notions of stopping time and predictable stopping time which corresponds quite closely to the intuitive ideas. This interpretation was first discovered by Galmarino (see Ito and M^c^Kean [1], p. 86) and later expanded and used systematically by Courrège and Priouret [1].

Let E be a Polish space [(1)] and let $\bar{E}$ be the space obtained by adjoining an isolated point ∂ to E. We use the term E-valued r.c.l.l. path with lifetime to denote any r.c.l.l. mapping ω of $\mathbb{R}_+$ into $\bar{E}$ with the following property 94

(94.1) The set $\{t : \omega(t) = \partial\}$ is a closed half-line $[a,+\infty[$

(a may equal 0 or $+\infty$). The number a is called the lifetime of ω and is denoted by $\zeta(\omega)$.

From an intuitive point of view, we speak of paths with values in E because ∂ is not a "true" state, it is a mathematical trick to work with E-valued paths which are not defined on the whole line.

We denote by Ω the set of all paths with lifetime and by $[\partial]$ the unique element

(1) The fact that E is Polish, rather than metrizable and separable, isn't used in the proofs below. See just remark 98 (e).

of Ω with zero lifetime. For all $\omega \in \Omega$ we set $\omega(t) = X_t(\omega)$ and introduce the σ-fields $\mathcal{F}^o = \sigma(X_s, s \geq 0)$ and $\mathcal{F}^o_t = \sigma(X_s, 0 \leq s \leq t)$. We are going to study the filtration $(\mathcal{F}^o_t)$, its stopping times, the optional and predictable σ-fields, etc...

If we wich to speak suitably about predictable processes, we must define $\mathcal{F}^o_{0-}$ there are two reasonable choices, either $\mathcal{F}^o_{0-} = \{\emptyset, \Omega\}$, $X_{0-} = \partial$, or $\mathcal{F}^o_{0-} = \mathcal{F}^o_0$, $X_{0-} = X_0$. For simplicity we choose the first definition.

95 We begin by defining some mappings of Ω into Ω which play a fundamental role here and elsewhere.

DEFINITION. *For all* $\omega \in \Omega$ *and all* $t \in \mathbb{R}_+$, *we define the elements* $\theta_t\omega$, $a_t\omega$ *and* $k_t\omega$ *of* Ω *by the following formulae*

(95.1) $$X_s(\theta_t\omega) = X_{s+t}(\omega) \qquad (s \in \mathbb{R}_+)$$

(95.2) $$X_s(a_t\omega) = X_{s\wedge t}(\omega)$$

(95.3) $$X_s(k_t\omega) = X_s(\omega) \text{ if } s < t, = \partial \text{ if } s \geq t.$$

These three operations are called translation by t, stopping at t and killing at t. *We also make the conventions*

(95.4) $$X_\infty(\omega) = \partial,\ k_0\omega = \theta_\infty\omega = [\partial],\ a_\infty\omega = k_\infty\omega = \omega.$$

Let ω *and* ω' *be two elements of* Ω *and* t *an element of* $\mathbb{R}_+$. *We denote by* $\omega/t/\omega'$ *the element of* Ω *defined by*

(95.5) $$\text{if } t \leq \zeta(\omega),\ X_s(\omega/t/\omega') = X_s(\omega) \text{ for } s < t, = X_{s-t}(\omega') \text{ for } s \geq t$$
$$\text{if } t > \zeta(\omega),\ \omega/t/\omega' = \omega.$$

The following theorem groups some simple properties of the operators just defined.

96 THEOREM. (a) *For all* s *and* t,

(96.1) $$\theta_s \circ \theta_t = \theta_{s+t},\ a_s \circ a_t = a_{s\wedge t},\ k_s \circ k_t = k_{s\wedge t}$$

(96.2) $$a_t \circ \theta_s = \theta_s \circ a_{t+s},\ k_t \circ \theta_s = \theta_s \circ k_{t+s}.$$

(b) *The mapping* $(t,\omega) \mapsto \theta_t\omega$, $a_t\omega$, $k_t\omega$ *of* $\mathbb{R}_+ \times \Omega$ *into* $(\Omega, \mathcal{F}^o)$ *are respectively : measurable relative to* $\mathcal{B}(\mathbb{R}_+) \times \mathcal{F}^o$, *optional relative to* $(\mathcal{F}^o_t)$ *and predictable relative to* $(\mathcal{F}^o_t)$ *(with the convention* $\mathcal{F}^o_{0-} = \{\emptyset, \Omega\}$*)*.

(c) *For all* t, $\mathcal{F}^o_t = a_t^{-1}(\mathcal{F}^o)$ *and* $\mathcal{F}^o_{t-} = k_t^{-1}(\mathcal{F}^o)$. *A* $\mathcal{F}^o$*-measurable r.v.* Z *is measurable with respect to* $\mathcal{F}^o_t$ *(resp.* $\mathcal{F}^o_{t-}$*), if and only if* $Z = Z \circ a_t$ *(resp.* $Z = Z \circ k_t$*)*.

(d) *The mapping* $(t,\omega,\omega') \mapsto \omega/t/\omega'$ *is measurable from* $\mathcal{B}(\mathbb{R}_+) \times \mathcal{F}^o \times \mathcal{F}^o$ *to* $\mathcal{F}^o$. *More precisely* $((t,\omega),\omega') \mapsto \omega/t/\omega'$ *is measurable from* $\mathcal{P} \times \mathcal{F}^o$ *to* $\mathcal{F}^o$.

Proof : (a) is obvious. To show (b), it is sufficient to verify that for every continuous function f on $\bar{E}$ and every $s \in \mathbb{R}_+$, the real-valued processes $(t,\omega) \mapsto f(X_s(\theta_t\omega)) = f(X_{s+t}(\omega))$, $f(X_s(a_t\omega)) = f(X_{s\wedge t}(\omega))$ and $f(X_s(k_t\omega))$ are respectively measurable, optional and predictable. The first one is obviously measurable and the second one optional because it is adapted and r.c.l.l. As for the third one, it can also be written as $f(X_s)I_{]\!]s,\infty[\![} + f(\partial).I_{[\![s,\infty[\![}$: it is therefore predictable.

The first part of (c) is immediate by the above. The second is easily verified in a direct way, but it also follows from I.18 (see 97). To show (d), one first verifies that the mapping $((t,\omega),\omega') \mapsto \omega/t/\omega'$ is $(\mathcal{B}(\mathbb{R}_+) \times \mathcal{F}^o) \times \mathcal{F}^o$-measurable, by proceeding as above. It is then noted that $\omega/t/\omega' = k_t\omega/t/\omega'$ identically and that the mapping $(t,\omega) \mapsto (t,k_t\omega)$ is a measurable mapping of $\mathcal{P}$ into $\mathcal{B}(\mathbb{R}_+) \times \mathcal{F}^o$.

We now characterize the optional and predictable σ-fields on $\mathbb{R}_+ \times \Omega$.

THEOREM. (a) The predictable σ-field $\mathcal{P}$ is generated by the two mappings $(t,\omega) \mapsto t$ and $(t,\omega) \mapsto k_t(\omega)$. The optional σ-field $\mathcal{O}$ is generated by the two mappings $(t,\omega) \mapsto t$ and $(t,\omega) \mapsto a_t\omega$ and also by the σ-field $\mathcal{P}$ and the mapping $(t,\omega) \mapsto X_t(\omega)$. 97

(b) Let (Z_t) be a measurable process. For (Z_t) to be predictable (resp. optional) it is necessary and sufficient that $Z_t = Z_t \circ k_t$ for all t (resp. $Z_t = Z_t \circ a_t$ for all t) or also that the process (Z_t) be adapted to the family $(\mathcal{F}^o_{t-})$ (resp. $\mathcal{F}^o_t$).

Proof : For simplicity we deal with the predictable case. We begin by noting that $\mathcal{F}^o_{t-}$, which is generated by the X_s such that $s < t$, is equal to $k_t^{-1}(\mathcal{F}^o)$, so that we have the equivalence

$$(Y \text{ is } \mathcal{F}^o_{t-}\text{-measurable}) \Leftrightarrow (\text{there exists a } \mathcal{F}^o\text{-measurable } Y' \text{ such that } Y = Y' \circ k_t).$$

But $k_t = k_t \circ k_t$ and hence

$$(Y \text{ is } \mathcal{F}^o_{t-}\text{-measurable}) \Leftrightarrow (Y = Y \circ k_t).$$

The equivalence of the two forms given in assertion (b) then follows immediately. We now consider the two following σ-fields on $\mathbb{R}_+ \times \Omega$

$\mathcal{H}$, generated by the mapping $(t,\omega) \mapsto (t,k_t\omega)$ with values in $(\mathbb{R}_+ \times \Omega, \mathcal{B}(\mathbb{R}_+) \times \mathcal{F}^o)$,

$\mathcal{K}$, generated by the measurable processes Z such that $Z_t = Z_t \circ k_t$ for all t (i.e. the processes adapted to the family $\mathcal{F}^o_{t-}$).

Obviously $\mathcal{H} \subset \mathcal{K}$ and none the less obviously $\mathcal{K} \subset \mathcal{H}$ (for if Z is $\mathcal{K}$-measurable, then $Z(t,\omega) = Z(t,k_t(\omega))$ by definition!). On the other hand, a predictable process is measurable and adapted to $(\mathcal{F}^o_{t-})$, hence $\mathcal{P} \subset \mathcal{K}$ and $\mathcal{H} \subset \mathcal{P}$ by 96.

For the optional case the argument is the same using a_t instead of k_t. There only remains one point to verify, namely that $\mathcal{O}$ is generated by $\mathcal{P}$ and the process (X_t). But for all $x \in E$ let $[x]$ be the constant path equal to x ; the mapping $x \mapsto [x]$ is measurable from E to $(\Omega, \mathcal{F}^o)$ and

$$a_t\omega = k_t\omega/t/[X_t(\omega)]$$

so that $(t,\omega) \mapsto a_t\omega$ is measurable with respect to the σ-field generated by $(t,\omega) \mapsto X_t(\omega)$ and $(t,\omega) \mapsto k_t(\omega)$ (96). This completes the proof.

98 REMARKS. (a) It follows in particular from (b) that every progressive process with respect to $(\mathcal{F}^o_t)$ is optional. This does not contradict the example in no. 91 (b) : in the latter case we were comparing the optional and progressive σ-fields of a canonical filtration augmented so as to satisfy the usual conditions.

The reader will be able to verify that, in the notation of (b), the process (Z_t) is progressive with respect to the family $(\mathcal{F}^o_{t+})$ if and only if $Z_s = Z_s \circ k_t$ (or $Z_s \circ a_t$) for every pair (s,t) such that $s < t$. There is no analogous characterization for optional processes relative to $(\mathcal{F}^o_{t+})$.

(b) If we dealt with right-continuous paths (instead of r.c.l.l. paths), the results concerning the σ-field $\mathcal{P}$ would be valid without modification, but it would be necessary to replace the σ-field $\mathcal{O}$ by the σ-field $\mathcal{O}'$ generated by the right-continuous adapted processes. It can be shown that for every law $\mathbb{P}$ every element of $\mathcal{O}'$ is $\mathbb{P}$-indistinguishable from an element of $\mathcal{O}$, but we have not given the proof in this volume. See Dellacherie-Meyer [2].

(c) When we speak of the σ-field generated on $\mathbb{R}_+ \times \Omega$ by $(t,\omega) \mapsto (t,k_t\omega)$, the first component is "almost" superfluous. For $t \wedge \zeta = \zeta \circ k_t$ identically, so that on the stochastic interval $[0,\zeta]$ - which is the interesting part of $\mathbb{R}_+ \times \Omega$, for nothing happens after ζ - the σ-field is generated by $(t,\omega) \mapsto k_t\omega$ alone. For a study of the σ-field generated by $(t,\omega) \mapsto \theta_t\omega$, see Azéma [1] - we also hope to find room for it in a later chapter. No one seems to have yet made a study of the σ-field generated by $(t,\omega) \mapsto (t,\theta_t\omega)$.

(d) Let Z be any $\mathcal{F}^o$-measurable r.v., and (Z_t) be the process $(Z \circ k_t)$. Then (Z_t) is a measurable process and $Z_t = Z_t \circ k_t$, so that (Z_t) is predictable according to 97 (b). We also have $Z_t(\omega) = Z(\omega)$ if $t \geq \zeta(\omega)$, since in this case $k_t\omega = \omega$.

Assume now we restrict Ω to be the set of all paths with a finite lifetime. Then the theory we have developed still applies, and the above result has a converse: any predictable process (Z_t) such that $Z_t = Z_\zeta$ for $t \geq \zeta$ (i.e. any predictable process which is stopped at time ζ) can be represented uniquely as $Z_t = Z \circ k_t$, where Z is a $\mathcal{F}^o$-mesurable r.v.. Uniqueness is obvious : since ζ is finite we may take $t = \zeta$ in the above equality, and get $Z_\zeta = Z \circ k_\zeta = Z$. On the other hand, if (Z_t) is predictable and stopped at ζ it is a trivial matter to check that $Z_\zeta \circ k_t = Z_t$ for all t.

This remark is due to Azéma. It is important, since it allows to reduce problems concerning processes to problems concerning random variables, whenever the space is provided with killing operators. Of course, we have considered here only a particular case, that of canonical spaces, but we hope to say more on this subject in later chapters.

(e) Theorem 97 was not discovered in this way : here is the original proof. Since the space E is Polish (we only used the metrizability of E in the above proofs), we know that the set Ω of E-valued r.c.l.l. mappings with lifetime is Lusin (19). Then consider the σ-field $\mathcal{D}$ generated by $\mathcal{P}$ and the process (X_t), the optional σ-field $\mathcal{O}$, the σ-field $\mathcal{A}$ generated by the measurable processes adapted to $(\mathcal{F}^0_t)$ and finally the σ-field $\mathcal{M} = \mathcal{B}(\mathbb{R}_+) \times \mathcal{F}^0$; then $\mathcal{D} \subset \mathcal{O} \subset \mathcal{A} \subset \mathcal{M}$ and $\mathcal{M}$ is a Blackwell σ-field ; on the other hand $\mathcal{D}$ is separable and the first three σ-fields have the same atoms : (s,ω) and (s',ω') belong to the same atom of $\mathcal{D}$ or $\mathcal{A}$ if and only if $s = s'$ and $a_s\omega = a_{s'}\omega'$. Then by Blackwell's theorem $\mathcal{D} = \mathcal{O} = \mathcal{A}$[1].

This proof may appear ridiculous alongside the elementary proof. However in some respects it is more natural than the proof of 97. We illustrate the method using Blackwell's theorem by a result which we leave to the reader.

Let $(\Omega, \mathcal{F})$ be a measurable space filtered by an increasing family $(\mathcal{F}^0_t)$ of Blackwell σ-fields. Then every stopping time of the family $(\mathcal{F}^0_{t-})$ is predictable.

We now state "Galmarino's test", which amounts to studying separately a very special case of 97 (b), that where (Z_t) is the indicator of a stochastic interval. Since we are not paying for the paper, we shall give the three statements with no "resp." and then some comments.

In the three statements, T is an $\mathcal{F}^0$-measurable function with values in $\overline{\mathbb{R}}_+$ and A an element of $\mathcal{F}^0$.

THEOREM. (a) T is a predictable time if and only if for every $t \in \mathbb{R}_+$[2] we have 99

(99.1) $(T(\omega) \le t,\ X_s(\omega) = X_s(\omega')$ for all $s < t) \Rightarrow (T(\omega) = T(\omega'))$.

(b) Then A belongs to $\mathcal{F}^0_{T-}$, if and only if

(99.2) $(\omega \in A,\ T(\omega) = T(\omega'),\ X_s(\omega) = X_s(\omega')$ for all $s < T(\omega)) \Rightarrow (\omega' \in A)$.

THEOREM. (a) T is a stopping time, if and only if for every $t \in \mathbb{R}_+$ we have 100

(100.1) $(T(\omega) \le t,\ X_s(\omega) = X_s(\omega')$ for all $s \le t) \Rightarrow (T(\omega) = T(\omega'))$.

(b) Then A belongs to $\mathcal{F}^0_T$, if and only if

(100.2) $(\omega \in A,\ T(\omega) = T(\omega'),\ X_s(\omega) = X_s(\omega')$ for all $s \le T(\omega)) \Rightarrow (\omega' \in A)$.

THEOREM. (a) T is a wide sense stopping time, if and only if for every $t \in \mathbb{R}_+$ we have 101

---

(1) The optional and progressive σ-fields with respect to $(\mathcal{F}^0_{t+})$ have the same atoms but are not separable : we can deduce nothing.
(2) Note that for $t = \infty$ the implication is obvious.

(101.1) $(T(\omega) \leq t$, there exists $\varepsilon > 0$ such that $X_s(\omega) = X_s(\omega')$ for $s \leq t + \varepsilon)$ $\Rightarrow (T(\omega) = T(\omega'))$.

(b) Then A belongs to $\mathcal{F}^0_{T+}$, if and only if

(101.2) $(\omega \in A, T(\omega) = T(\omega')$, there exists $\varepsilon > 0$ such that $X_s(\omega) = X_s(\omega')$ for $s \leq T(\omega) + \varepsilon) \Rightarrow (\omega' \in A)$.

Proof : We begin with the three assertions (a) and first deal with the second one which is simpler. We first note that if implication (100.1) holds, then we have the weaker implication

(100.1') $(T(\omega) \leq t, X_s(\omega) = X_s(\omega')$ for all $s \leq t) \Rightarrow (T(\omega') \leq t)$.

But conversely, if (100.1') holds for all t, then (100.1) holds. Suppose indeed that the condition on the left-hand side of (100.1') is satisfied and let $r = T(\omega)$, so that $r \leq t$. The condition then implies that

$$T(\omega) \leq r,\ X_s(\omega) = X_s(\omega') \text{ for all } s \leq r$$

which (by (100.1') with r instead of t) implies $T(\omega') \leq r$. In other words, the left-hand side of (100.1') implies $T(\omega') \leq T(\omega)$. Since it also implies the right-hand side, exchanging ω and ω' we have the relation

$$T(\omega') \leq t,\ X_s(\omega) = X_s(\omega') \text{ for } s \leq t$$

and this implies $T(\omega) \leq T(\omega')$. Hence finally $T(\omega) = T(\omega')$, that is (100.1).

Now, what does (100.1') mean ? The relation $(X_s(\omega) = X_s(\omega')$ for $s \leq t)$ can be written simply $(a_t\omega = a_t\omega')$, so that (100.1') means that the set $\{T \leq t\}$ is saturated with respect to the equivalence relation $R_t : (a_t\omega = a_t\omega')$. Since ω and $a_t\omega$ are equivalent mod. R_t, this amounts to saying that $\{T \leq t\} = a_t^{-1}\{T \leq t\}$, which using 96 (c) and the fact that T is $\mathcal{F}^0$-measurable by hypothesis, means exactly that $\{T \leq t\} \in \mathcal{F}^0_t$ or that T is a stopping time.

We shall not give the details for the third statement : we get the condition $\{T \leq t\} = a^{-1}_{t+\varepsilon}\{T \leq t\}$ for all $\varepsilon > 0$ or $\{T \leq t\} \in \mathcal{F}^0_{t+}$.

Similarly, for the first statement, we get the equivalence of (99.1) to the condition $\{T \leq t\} \in \mathcal{F}^0_{t-}$ for all t. But here there is something of interest : we have at our disposal no general theorem stating that, if this condition is satisfied, then T is a predictable stopping time, and we must use Theorem 97 (b) : the measurable process (Z_t), indicator of $[\![T,\infty[\![$, satisfies the relations $Z_t = Z_t \circ k_t$ and is therefore predictable.

We now pass to assertions (b) and deal for example with the first statement. Suppose that $A \in \mathcal{F}^0_{T-}$ and that ω and ω' satisfy the left-hand side of (99.2). We set $t = T(\omega)$. If $t = +\infty$ there is nothing to prove.

Otherwise, we remark that $T_A(\omega) \leq t$, $X_s(\omega) = X_s(\omega')$ for $s \leq t$; since T_A is predictable stopping time, we deduce that $T_A(\omega') = T_A(\omega)$, hence $T_A(\omega') < \infty$ and finally

$\omega' \in A$.

Conversely, suppose that T is a predictable stopping time and that implication (99.2) is true. We verify that that T_A is a predictable stopping time using (a) : suppose that

$$T_A(\omega) \leq t,\ X_s(\omega) = X_s(\omega') \quad \text{for } s < t.$$

We deduce from it the left-hand side of (99.1) and hence, T being predictable, that $T(\omega) = T(\omega')$: but then the left-hand side of (99.2) is true, hence $\omega' \in A$ and $T_A(\omega) = T_A(\omega')$, the required result.

The arguments for the other two parts (b) are exactly similar.

REMARKS. (a) One shouldn't live with too many illusions about the practical value of the "test" : in order to show that a random variable T is a stopping time, the most difficult step in general consists in verifying that it is $\mathcal{F}^o$-measurable (and moreover it is rarely so, one must usually complete the σ-fields), so that the "test" can only be applied after most of the work has been done. We have indicated it mainly because of the light it throws on the very notion of a stopping time. 102

(b) Let T be an $\mathcal{F}^o$-measurable function with values in $\bar{\mathbb{R}}$. It follows immediately from the "test" that

T is a stopping time if and only if $\{T = t\} \in \mathcal{F}^o_t$ for all t

T is a predictable time if and only if $\{T = t\} \in \mathcal{F}^o_{t-}$ for all t.

In other words, we have the same characterizations as in the discrete case (49, 68) but only on a very special kind of filtered space.

By way of illustration of "Galmarino's test", and also of the symbols $\omega/t/w$ which we have scarcely used, we present a special case of a beautiful theorem on the decomposition of stopping times due to Courrège and Priouret [1].

THEOREM. _Let_ S _and_ T _be two stopping times of the family_ $(\mathcal{F}^o_t)$ _such that_ $S \leq T$. _Then there exists a function_ $U(\omega,w)$ _on_ $\Omega \times \Omega$ _with values in_ $\bar{\mathbb{R}}$ _with the following properties_ 103

(1) $U(.,.)$ is $\mathcal{F}^o_S \times \mathcal{F}^o$_-measurable_ ;

(2) $U(\omega,w) = 0$ _if_ $S(\omega) = +\infty$ _or if_ $S(\omega) < \infty$ _and_ $X_0(w) \neq X_S(\omega)$;

(3) _for all_ ω, $U(\omega,.)$ _is a stopping time_ ;

(4) $T(\omega) = S(\omega) + U(\omega,\theta_S\omega)$ _for all_ ω.

Proof : We define $W(\omega) = T(\omega) - S(\omega)$ with the convention $\infty - \infty = 0$; W is positive and $\mathcal{F}^o$-measurable. We then set $V(\omega,t,w) = W(\omega/t/w)$, a $(\mathcal{F}^o \times \mathcal{B}(\mathbb{R}_+) \times \mathcal{F}^o)$-measurable function (96 (d)). Finally we set

$$U(\omega,w) = V(\omega,S(\omega),w) \text{ if } S(\omega) < \infty \text{ and } X_0(w) = X_S(\omega)$$
$$= 0 \text{ otherwise}$$

and verify that U satisfies the statement : (2) is part of the definition ; (4) is obvious if $S(\omega) = +\infty$, and if $S(\omega) < \infty$ it follows from the equalities

$$U(\omega,\theta_S\omega) = V(\omega,S(\omega),\theta_S\omega) \text{ for } X_0(\theta_S\omega) = X_S(\omega)$$
$$= W(\omega/S(\omega)/\theta_S\omega) = W(\omega) = T(\omega) - S(\omega).$$

To verify (1), we first note - by a simple composition of applications - that $U(.,.)$ is $\mathcal{F}^0 \times \mathcal{F}^0$-measurable. Then S a strict sense stopping time, so that $U(\omega,w) = U(a_S\omega,w)$ for all w, and finally a_S is a measurable from $\mathcal{F}^0_S$ into $\mathcal{F}^0$.

There remains to verify (3), which may be amusing since it uses the "test". If $S(\omega) = +\infty$, there is nothing to prove. So we suppose that $S(\omega) = s < \infty$ and show that

$$(U(\omega,w) \le t,\ X_r(w) = X_r(w') \text{ for } r \le t) \Rightarrow (U(\omega,w') = U(\omega,w)).$$

The relation $X_r(w) = X_r(w')$ for $r \le t$ implies in particular that $X_0(w) = X_0(w')$. If $X_0(w) \neq X_S(\omega)$, we have $U(\omega,w) = U(\omega,w') = 0$ and the property is satisfied. Suppose then that $X_0(w) = X_0(w') = X_S(\omega)$; we must show that $V(\omega,S(\omega),w) = V(\omega,S(\omega),w')$ or equivalently, since $S(\omega) = s$,

$$T(\omega/s/w) - S(\omega/s/w) = T(\omega/s/w') - S(\omega/s/w').$$

As $X_s(\omega) = X_0(w) = X_0(w')$, we have $X_r(\omega/s/w) = X_r(\omega/s/w') = X_r(\omega)$ for $r \le s$ and hence, S being a stopping time, $S(\omega/s/w) = S(\omega/s/w') = S(\omega) = s$. Then the relation $X_r(w) = X_r(w')$ for $r \le t$ implies that $X_r(\omega/s/w) = X_r(\omega/s/w')$ for $r \le s + t$; as T is a stopping time and $T(\omega/s/w) = S(\omega/s/w) + U(\omega,w) \le s + t$, we deduce that $T(\omega/s/w) = T(\omega/s/w')$ and hence also that $U(\omega,w) = U(\omega,w')$.

Study of an example

This example is studied in Dellacherie [1], but the proof contains some gaps (as pointed out to us by R. Getoor) and we hope that we have corrected them here. It may look trivial, but still, it has some importance in concrete applications of probability (cf. Chou-Meyer [1]). We shall return to it in a later chapter to illustrate martingale theory.

104 The set Ω here is the interval $]0,\infty[$ with its Borel σ-field, denoted by $\mathcal{F}^0$, and we use T to denote the identity mapping of Ω into $\bar{\mathbb{R}}_+$.

For every $t \in \bar{\mathbb{R}}_+$, let $\mathcal{F}^0_t$ be the σ-field generated by $T \wedge t$ or equivalently by the Borel sets of $]0,t[$ and the atom $[t,\infty[$ (1). The family $(\mathcal{F}^0_t)$ is a filtration which satisfies $\mathcal{F}^0_t = \mathcal{F}^0_{t-}$ but not $\mathcal{F}^0_t = \mathcal{F}^0_{t+}$ ($\mathcal{F}^0_{t+}$ is obtained by splitting the atom $[t,\infty[$ into two atoms, $\{t\}$ and $]t,\infty[$). The random variable T is a stopping time of $(\mathcal{F}^0_{t+})$ but not of $(\mathcal{F}^0_t)$.

There is a very simple characterization of the stopping times of $(\mathcal{F}^0_{t+})$ and of those of $(\mathcal{F}^0_t)$:

(1) We make the convention $\mathcal{F}^0_{0-} = \mathcal{F}^0_0 = \{\emptyset,\Omega\}$.

THEOREM. A positive variable S is a stopping time of $(\mathcal{F}_t^o)$ (resp. $(\mathcal{F}_{t+}^o)$) if and only if there exists a constant s (possibly equal to $+\infty$) such that 105

(a) $S > T$ on $\{T < s\}$ (resp. $S \geq T$ on $\{T \leq s\}$) ;

(b) $S = s$ on $\{T \geq s\}$ (resp. $S = s$ on $\{T > s\}$).

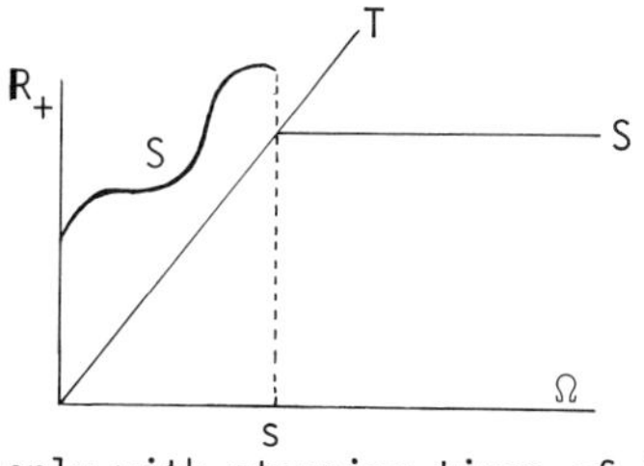

Proof : We shall deal only with stopping times of $(\mathcal{F}_t^o)$. If S satisfies the conditions of the theorem, then $\{S \leq t\} \in \mathcal{F}_t^o$ for all t. Indeed, if $t < s$, the set $\{S \leq t\}$ is contained in $\{T < t\}$, it is a Borel set of $]0,t[$ and belongs to $\mathcal{F}_t^o$; if $t \geq s$, the set $\{S \leq t\}$ contains the whole of the atom $[t,\infty[$ and it also belongs to $\mathcal{F}_t^o$.

Conversely, let S be a stopping time of $(\mathcal{F}_t^o)$. If $S > T$ everywhere, there is nothing to prove. If there exists some ω such that $S(\omega) \leq T(\omega)$, we take $s = S(\omega) \leq T(\omega) = \omega$. The set $\{S = s\}$ belongs to $\mathcal{F}_s^o$ and contains ω, which belongs to the atom $[s,\infty[$; hence it contains the whole of the atom $[s,\infty[$, thus giving condition (b). Now let $\omega' < s$. If it were true that $s' = S(\omega') \leq T(\omega') = \omega'$, it would follow that $s' < s$. The set $\{S = s'\} \in \mathcal{F}_{s'}^o$ would contain the point ω' of the atom $[s',\infty[$, hence the whole of the atom, and we would have $S = s'$ on $[s,\infty[$ whereas we have $S = s$ there and $s \neq s'$. Consequently, $S(\omega') > T(\omega')$ and condition (b) is proved.

THEOREM. Every stopping time of $(\mathcal{F}_t^o)$ is predictable. 106

Proof : Let S be a stopping time of $(\mathcal{F}_t^o)$ and s the constant associated with it by 105. If $s = 0$, then $S \equiv 0$ and hence S is predictable ; suppose that $s > 0$ and let (s_n) be a sequence of positive reals such that $s_n \uparrow\uparrow s$. Then S is foretold by the sequence (S_n) of stopping times of $(\mathcal{F}_t^o)$ defined by

$$S_n = n \wedge ((1 - \frac{1}{n})S + \frac{T}{n}) \text{ on } \{T < n \wedge s_n\}$$

$$S_n = n \wedge s_n \text{ on } \{T \geq n \wedge s_n\}.$$

We now give $(\Omega,\mathcal{F}^o)$ a probability law $\mathbb{P}$. We denote by $\mathcal{F}$ the completed σ-field of $\mathcal{F}^o$ and by $(\mathcal{F}_t)$ the usual augmentation of $(\mathcal{F}_t^o)$.

THEOREM. If the law $\mathbb{P}$ is diffuse, the filtration $(\mathcal{F}_t)$ is quasi-left-continuous and T is totally inaccessible. 107

If the law $\mathbb{P}$ is purely atomic and non-degenerate, T is a non-predictable accessible time.

Proof : We begin with the second assertion. The law $\mathbb{P}$ is carried by a countable set

D and the graph of T is then P-a.s. contained in the union of the graphs $[\![t]\!]$, $t \in D$. Since constants are predictable times, T is accessible (81). Suppose that $\mathbb{P}$ is non-degenerate. Then P is non-zero at two distinct points u and v such that $u < v$. If T were predictable, there would exist a predictable time S of $(\mathcal{F}_t^o)$ such that $S = T$ a.s. and hence $S(u) = T(u)$ and $S(v) = T(v)$. By 105 this implies $u = v$, which is absurd.

Suppose that $\mathbb{P}$ is diffuse. Since every set of the form $\{t\}$ is negligible, the σ-fields obtained by adjoining all the negligible sets to $\mathcal{F}_{t-}^o = \mathcal{F}_t^o$ and to $\mathcal{F}_{t+}^o$ are equal, so that $(\mathcal{F}_t)$ is simply the completed filtration of $(\mathcal{F}_t^o)$. We first show that T is totally inaccessible. Let S be a stopping time of the family $(\mathcal{F}_t)$ such that $S \leq T$. Then there exits a stopping time R of the family $(\mathcal{F}_{t+}^o)$ such that $S = R$. a.s. (59) and, replacing R by $R \wedge T$ if necessary, we can assume that $R \leq T$ everywhere. But then, by 105, R is of the form $T \wedge s$ and $S = T \wedge s$ a.s. Now let (S_n) be an increasing sequence of stopping times bounded above by T ; each one is a.s. of the form $T \wedge s_n$, the s_n increase to a number s and the set $\{\lim_n S_n = T, S_n < T \text{ for all } n\}$ is a.s. contained in $\{T = s\}$, which is negligible because P is diffuse.

Finally we show that the family $(\mathcal{F}_t)$ is quasi-left-continuous. Let U be a predictable stopping time of $(\mathcal{F}_t)$ and A be an element of $\mathcal{F}_U$; we show that A belongs to $\mathcal{F}_{U-}$. Since all the negligible sets belong to $\mathcal{F}_{0-}$, it is sufficient to show that A is a.s. equal to an element of $\mathcal{F}_{U-}$. But U_A is a stopping time of $(\mathcal{F}_t)$ and hence is equal a.s. to a stopping time V of $(\mathcal{F}_{t+}^o)$ (59). As U is predictable and T totally inaccessible, $\mathbb{P}\{U = T\} = 0$ and hence, replacing V by $V_{\{V \neq T\}}$ if necessary, we can assume that V is nowhere equal to T. Then we see by 105 that V is a stopping time of $(\mathcal{F}_t^o)$ and hence a predictable time and A is a.s. equal to the set $(A \cap \{U = \infty\}) \cup \{V = U\}$, which belongs to $\mathcal{F}_{U-}$.

108 REMARKS. (a) If $\mathbb{P}$ is diffuse, the completed filtration $(\mathcal{F}_t)$ satisfies $\mathcal{F}_{t-} = \mathcal{F}_t = \mathcal{F}_{t+}$ for all t, so that every stopping time of $(\mathcal{F}_t)$ still is a stopping time of $(\mathcal{F}_{t-})$ - without implying that every stopping time of $(\mathcal{F}_t)$ is predictable.

(b) Suppose that P is diffuse and that the greatest lower bound of the support of P is 0. We have seen that every stopping time $S \leq T$ is a.s. equal to $T \wedge s$, where s is a constant. Hence $S < T$ a.s. only if $S = 0$ a.s. In other words, the optional set $]\!]0,T[\![$ does not possess a complete, or even almost-complete, optional cross-section.

APPENDIX TO CHAPTER III

The results below aren't less interesting than those of the main text : they just lack (for the moment) important applications either to measure theory or to potential theory. One should bear in mind that the results of Chapter III, or even those in Chapter IX of Bourbaki's *General Topology* are only the lower stages of the descriptive theory of sets, developed by Polish and Russian mathematicians and then by modern logicians to incredible heights. The last results in the appendix indicate the next stage above.

The numbering follows that of Chapter III.

Souslin schemes

In probability theory analytic sets appear naturally as projections of Borel sets, whence the definition which we have adopted in the text. But the oldest (and still most used) definition is that of Souslin. We prove it here to be equivalent to the definition in the text, and to that given in the first edition of our book.

We denote by S the set of finite sequences of integers and by Σ the set $\mathbb{N}^{\mathbb{N}}$ of infinite sequences of integers : if $\mathbb{N}$ is given the discrete topology and Σ the product topology, Σ is a Polish space and a $\mathcal{K}_{\delta}$ of the compact metrizable space $(\mathbb{N} \cup \{\infty\})^{\mathbb{N}}$. We denote by $|s|$ the length of a finite sequence $s \in S$. The notation $s < t$, where $s \in S$ and $t \in S$ or $t \in \Sigma$, means that t begins with s : for example s = 3,1,4 and t = 3,1,4,1,6. Finally, for $\sigma \in \Sigma$, we denote the n-th term of σ by $\sigma(n)$ and the finite sequence $\sigma(1),\ldots,\sigma(n)$ by $\sigma|n$, with analogous notation s(n) and $s|n$ if $s \in S$ and $n \leq |s|$. 75

DEFINITION. *Let* $(F,\mathfrak{F})$ *be a paved space. A* Souslin scheme on $\mathfrak{F}$ *is a mapping* $s \mapsto B_s$ *of* S *into* $\mathfrak{F}$. *The* kernel *of the Souslin sheme is the set*

$$B = \bigcup_{\sigma} \bigcap_{s<\sigma} B_s = \bigcup_{\sigma} \bigcap_{n} B_{\sigma|n}.$$

We say also that B is the result of the *Souslin operation* (A) applied to the determining system $(B_s)_{s \in S}$. The Souslin scheme is said to be *regular* if $B_s \supset B_t$ for $s < t$. If $\mathfrak{F}$ is closed under ($\cap$f), the regular Souslin scheme $B'_s = \bigcap_{r<s} B_r$ is also a scheme on $\mathfrak{F}$ and still has the same kernel B ; hence there is no loss of generality in considering only regular schemes in this case.

We first show that every $\mathfrak{F}$-analytic set, even every $\mathfrak{F}$-analytic set in a slightly more general sense (see the first edition of this book, page 34), is the kernel of a Souslin scheme on $\mathfrak{F}$.

THEOREM. *Let* $(E,\mathcal{E})$ *and* $(F,\mathfrak{F})$ *be two paved sets, the paving* $\mathcal{E}$ *being semicompact. Let* B *be an element of* $(\mathcal{E} \times \mathfrak{F})_{\sigma\delta}$ *and* A *be its projection onto* F. *Then* A *is the Kernel of* 76 a Souslin scheme on $\mathcal{F}$.

Proof : By definition, B can be written in the form

$$B = \bigcap_n \bigcup_m (E_{nm} \times F_{nm}),\ E_{nm} \in \mathcal{E},\ F_{nm} \in \mathcal{F}.$$

We permute the operations of union and intersection, getting

$$B = \bigcup_\sigma \bigcap_n (E_{n\sigma(n)} \times F_{n\sigma(n)}).$$

Then for every $s \in S$ of length $|s| = n$ we write $A_s = F_{ns(n)}$ if $\bigcap_{m\leq n} E_{ms(m)} \neq \emptyset$ and $A_s = \emptyset$ otherwise, π denoting projection onto F, we have

$$\pi(\bigcap_{1\leq n\leq N} E_{n\sigma(n)} \times F_{n\sigma(n)}) = \bigcap_{1\leq n\leq N} A_{\sigma|n}.$$

Since the paving $\mathcal{E}$ is semicompact, by 6

$$\pi(\bigcap_n E_{n\sigma(n)} \times F_{n\sigma(n)}) = \bigcap_n A_{\sigma|n}$$

and taking the union over σ we get the required result

$$A = \pi(B) = \bigcup_\sigma \bigcap_n A_{\sigma|n}.$$

77 We pass to the converse. We even establish a stronger result : every kernel A of a Souslin scheme (A_s) on $\mathcal{F}$ is the projection of an element of $(\mathcal{E} \times \mathcal{F})_{\sigma\delta}$, where $\mathcal{E}$ is the paving consisting of all compact subsets in the compact metrizable space $E = (\mathbb{N} \cup \{\infty\})^{\mathbb{N}}$. So there is no need in definition 7 to take "all" compact metrizable spaces as auxiliary sets ; E is sufficient (1).

We introduce a convenient terminology : for all $s \in S$, the set $I_s = \{\sigma \in \Sigma : s < \sigma\}$ is called the islet of index s. Clearly $I_s \supset I_t$ if $s < t$, $I_s \subset I_t$ if $t < s$ and $I_s \cap I_t = \emptyset$ if s and t are not comparable. Islets are both open and closed and form a countable base of the topology of Σ.

THEOREM. Let A be the kernel of a Soulin scheme (A_s) on $\mathcal{F}$. Then A is the projection onto F of the following subset of $\Sigma \times F$

$$B = \bigcap_n \bigcup_{|s|=n} (I_s \times A_s) = \bigcup_\sigma \bigcap_n (I_{\sigma|n} \times A_{\sigma|n}) = \bigcup_\sigma [\{\sigma\} \times (\bigcap_n A_{\sigma|n})].$$

If $\Sigma \times F$ is considered as a subset of $E \times F$, B belongs to $(\mathcal{E} \times \mathcal{F})_{\sigma\delta}$.

Proof : The equality of the first and second expressions of B follows from the remarks on islets preceding the statement ; the equality of the second and third expressions is obvious. It follows immediately from the third expression that $\pi(B) = A$.

---

(1) The possibility of using a unique auxiliary compact set is not a great discovery, since all compact metrizable spaces can be imbedded in $[0,1]^{\mathbb{N}}$.

Let $\bar{I}_s$ be the closure of I_s in E ; since I_s is closed, $I_s = \bar{I}_s \cap \Sigma$ and $\bigcap_n \bigcup_{|s|=n} (I_s \times A_s)$ is the intersection of $\bigcap_n \bigcup_{|s|=n} (\bar{I}_s \times A_s)$ - which belongs to $(\mathcal{E} \times \mathcal{F})_{\sigma\delta}$ - with $\Sigma \times U$, where U is the union of all the A_s. As Σ is a $\mathcal{G}_\delta$ in E, Σ is also a $\mathcal{E}_{\sigma\delta}$; similarly U is a $\mathcal{F}_\sigma$ and hence a $\mathcal{F}_{\sigma\delta}$. Finaly, $\Sigma \times U$ is a $(\mathcal{E} \times \mathcal{F})_{\sigma\delta}$ and the theorem follows.

REMARK. We have seen that the $\mathcal{F}$-analytic sets are exactly those which are constructed from the Souslin operation (A). The result of 10 according to which $\mathcal{A}(\mathcal{A}(\mathcal{F})) = \mathcal{A}(\mathcal{F})$, therefore means that the kernel of a Souslin scheme on $\mathcal{A}(\mathcal{F})$ (which is itself the set of all kernels of Souslin schemes on $\mathcal{F}$) still is the kernel of some Souslin scheme on $\mathcal{F}$. The direct proof of this result (known as the "idempotency" of the operation (A)) requires a certain amount of ingenuity : see for example Hausdorff's Mengenlehre [1].

Supplement on Souslin and Lusin spaces

In no. 16 we defined a metrizable Souslin (resp. Lusin) space as any space homeomorphic to an analytic (resp. Borel) subspace of a compact metrizable space ; in no. 67 we extended this definition by saying that a Hausdorff space which is not necessarily metrizable is Souslin (resp. Lusin) if it is the continuous (resp. injective continuous) image of some metrizable Souslin (resp. Lusin) space. We now show that every analytic (resp. Borel) subset of a compact metrizable space is the continuous (resp. injective continuous) image of a Polish space. This will imply the equivalence between our definition and Bourbaki's (General Topology, Chapter IX, §6).

THEOREM. Let E be an analytic subset of a compact metric space F. Then there exists a continuous mapping of the Polish space $\Sigma = \mathbb{N}^{\mathbb{N}}$ onto E. 78

Proof : Let $\mathcal{F}$ denote the paving consisting of the compact subsets of F and let G be an auxiliary compact space, given the paving $\mathcal{G}$ of its compact subsets, such that E is the projection onto F of a set $H \in (\mathcal{G} \times \mathcal{F})_{\sigma\delta}$

$$H = \bigcap_n \bigcup_m (A_{nm} \times B_{nm}),\ A_{nm} \in \mathcal{G},\ B_{nm} \in \mathcal{F}.$$

Let d be a distance defining the topology of F. Replacing the B_{nm} if necessary by their intersections with suitable balls and modifying the indices m, we can suppose that all the compact sets B_{nm} have diameter (with respect to d) at most 1/n. We then transform this representation as in no. 76 : we get a representation of E as the kernel of a Souslin scheme (E_s) on $\mathcal{F}$ such that for all $s \in S$ the diameter of E_s is at most $1/|s|$. Bu then for all $\sigma \in \Sigma$ the intersection $E_\sigma = \bigcap_n E_{\sigma|n}$ either reduces to a single point or is empty. In the first case we denote the unique element of E_σ by $f(\sigma)$ and in the second case we set $f(\sigma) = x_0$, an arbitrarily chosen element of E. Clearly f is a mapping of Σ onto E and we leave to the reader the task of showing that it is continuous.

79 THEOREM. Let E be a Borel subset of a compact metrizable space. Then E is the injective continuous image of a closed (and hence Polish) subspace P of Σ.

Proof : We can suppose that E is imbedded in $[0,1/2]^{\mathbb{N}}$; by imbedding the latter set in $F = [0,1[^{\mathbb{N}}$, the problem is reduced to establishing the theorem for all Borel sets of F. We shall use an argument by monotone classes.

Let $A \subset F$; we define that $A \in \mathcal{P}$ if there exist a closed set P of Σ and an injective continuous mapping f of P onto A, and show that $\mathcal{P}$ has good closure properties and is sufficiently rich.

(a) $\mathcal{P}$ is closed under finite or countable intersections. We consider the countably infinite case : let A_n be elements of $\mathcal{P}$ which are images of closed sets P_n of Σ under continuous bijections f_n. We identify F with the diagonal Δ of $F^{\mathbb{N}}$ using the diagonal mapping $x \mapsto (x,x,\ldots)$ and denote by ϕ the mapping $\prod_n f_n$ of $\Sigma^{\mathbb{N}}$ into $F^{\mathbb{N}}$ and by P the closed subset $\phi^{-1}(F) \cap \prod_n P_n$ of $\Sigma^{\mathbb{N}}$. Clearly $f = \phi|_P$ is a continuous bijection of P onto $\bigcap_n A_n \subset F = \Delta$. Now P is a closed subset of $\Sigma^{\mathbb{N}}$, not of Σ ; it matters little since $\Sigma = \mathbb{N}^{\mathbb{N}}$, so $\Sigma^{\mathbb{N}} = \mathbb{N}^{\mathbb{N}\times\mathbb{N}}$ is homeomorphic to Σ, and P can be considered as a subset of Σ by transport of structure.

(b) The union of a finite or countable sequence of disjoint elements A_n of $\mathcal{P}$ belongs to $\mathcal{P}$. We again consider the countably infinite case and define the P_n and f_n as above. We can immediately define a continuous bijection f of the topological sum P of the P_n onto $\bigcup_n A_n$. Now P is a closed subset of the topological sum $\mathbb{N} \times \Sigma$ of a sequence of copies of Σ , and as above this is homeomorphic to Σ .

(c) Let $\mathcal{C}$ be the family of all subsets of F of the form $\prod_n I_n$, where for all n I_n is an interval of the form $[a_n,b_n[$ and $I_n = [0,1[$ except for a finite number of values of n. Note that $\mathcal{C}$ contains F and is closed under finite intersections and that the complement of an element of $\mathcal{C}$ is a finite union of elements of $\mathcal{C}$. We show that $\mathcal{C} \subset \mathcal{P}$. Since $\Sigma^{\mathbb{N}}$ is homeomorphic to Σ, it is obviously sufficient to show that there exists a continuous bijection of Σ onto each of the I_n or simply onto $[0,1[$. Such a bijection is given by

$$f(\sigma) = 1- 2^{-\sigma(1)} - 2^{-[\sigma(1)+\sigma(2)]} - 2^{-[\sigma(1)+\sigma(2)+\sigma(3)]} \ldots \quad (\sigma \in \Sigma = \mathbb{N}^{\mathbb{N}})$$

(recall that $\mathbb{N} = \{1,2,\ldots\}$, so that all the $\sigma(i)$ are > 0).

We can now complete the proof. Let $\mathcal{M}$ be the set of $A \subset F$ such that A and A^c belong to $\mathcal{P}$; we show that $\mathcal{M}$ is closed under $(\cup c, \cap c)$. It is sufficient to prove that for every sequence (A_n) of elements of $\mathcal{M}$, $\bigcap_n A_n$ and $\bigcup_n A_n$ belong to $\mathcal{P}$. For intersections this follows from (a) and for unions from (b) and the fact that $\bigcup_n A_n$ is also the union of the disjoint sets $B_1 = A_1$, $B_2 = A_2 \cap A_1^c$, $B_3 = A_3 \cap A_1^c \cap A_2^c, \ldots$, which belong to $\mathcal{P}$ by (a). We conclude by noting that $\mathcal{M}$ contains $\mathcal{C}$ by (c) and that $\mathcal{C}$ generates the Borel σ-field.

From the topological point of view, a metrizable Lusin space is homeomorphic to a Borel subspace of a compact metrizable space. From the measure theoretic point of view, the situation is much simpler : we now establish (following Christensen's [1] elegant proof) the following result due to Kuratowski :

THEOREM. All uncountable Lusin measurable spaces are isomorphic (in particular to 80
$[0,1]$, $\sum$ or $\{0,1\}^{\mathbb{N}}$).

Proof : Let L be an uncountable Lusin metrizable space. We compare it to the set $C = \{0,1\}^{\mathbb{N}}$ (homeomorphic to the Cantor set) by first constructing a measurable isomorphism f of L onto a Borel subset of C and then a measurable isomorphism g of C onto a Borel subset of L. After that we use the classical method of V. Bernstein to construct a measurable isomorphism of L onto C.

(a) Let (A_n) be a countable Boolean algebra generating the Borel σ-field of L. The mapping $f : x \mapsto (I_{A_n}(x))_{n \in \mathbb{N}}$ then is a measurable isomorphism of L onto a subset of C with its Borel σ-field (cf. I.11, an analogous result). The image f(L) is Borel by the Souslin-Lusin Theorem (21), but in fact it is interesting to prove this without 21 which appeals to the theory of analytic sets : since f is a measurable isomorphism, f(L) is Borel by the isomorphism extension theorem III.18-19.

(b) In the other direction, we consider a Polish space P and an injective continuous mapping h of P onto L. We construct an injective continuous mapping j of C into P and that will suffice : for $h \circ j = g$ will be an injective continuous mapping of the compact space C into the Hausdorff space L and will therefore be a homeomorphism of C onto a compact subset of L.

We give P the structure of a complete separable metric space. Let M be the set of all $x \in P$ such that every neighbourhood of x is uncountable ; M is closed with no isolated points and, since the topology of P has a countable base, $P\backslash M$ is countable and hence M is non-empty. Let x and y be two distinct points of M and M_0 and M_1 be two disjoint closed neighbourhoods of x and y, of diameter ≤ 1. It follows from the definition of M that M_0 and M_1 are not countable. Repeating the operation, we construct two disjoint non-countable closed subsets M_{00} and M_{01} of M_0 of diameter $\leq 1/2$ and similarly M_{10} and M_{11} in M_1. Following this procedure we can construct for every finite dyadic word s of length $|s| = n$ a non-countable closed set M_s of diameter $\leq 2^{1-n}$, such that

$$s \prec t \Rightarrow M_s \supset M_t \ ; \quad |s| = |t| \ , s \neq t \Rightarrow M_s \cap M_t = \emptyset.$$

The mapping which associates with every infinite sequence $\sigma \in C$ the unique element $j(\sigma)$ of $\bigcap_n M_{\sigma|n}$ is then continuous and injective.

(c) To simplify notation, we identify C with a (compact) subset of L, so that g is now the canonical injection of C into L and f is a measurable isomorphism of

L onto a Borel subset of C. We define

$$A_1 = L\setminus C,\ A_2 = f(A_1),\dots,\ A_{n+1} = f(A_n)$$

$$A = \bigcup_n A_n,\ B = L\setminus A$$

$$h(x) = f(x) \text{ if } x \in A,\ h(x) = x \text{ if } x \in B.$$

Since f maps L into C, A_1 and A_2 are disjoint and it is immediately verified that all the A_i are disjoint, that $B \subset C$ and that $f(A) = C\setminus B$. On the other hand, f induces an isomorphism of A_i onto A_{i+1}, hence h induces an isomorphism of A onto $f(A) = C\setminus B$ and an isomorphism of B onto itself, and finally h is an isomorphism of L onto C.

A cross-section theorem.

81 This is a nice cross-section theorem which is stronger than 44 (b) - it makes no use of the measure on E -and whose proof is simple. We take it from the book [1] by Hoffmann-Jørgensen but it is much older (Jankov 1941, Von Neumann 1949).

Recall that $\sum$ can be given a total ordering (denoted by $\le$), namely the lexicographic ordering defined as follows : let σ and τ be two elements of $\sum$, then $\sigma \le \tau$ if $\sigma = \tau$ or $\sigma(n) < \tau(n)$, where n denotes the smallest integrer i such that $\sigma(i) \ne \tau(i)$. Every non-empty closed subset of $\sum$ has a smallest element with respect to this ordering. For all $\tau \in \sum$ let $J_\tau = \{\sigma : \sigma \le \tau,\ \sigma \ne \tau\}$; it is not difficult to show that J_τ is open. On the other hand, the Borel σ-field of $\sum$ is generated by the J_τ. Let $s \in S$ be of length k, and let τ_n be the infinite sequence which coincides with s up to the k-th term and has the value n thereafter ; we write $L_s = \bigcup_n J_{\tau_n}$, which belongs to the σ-field generated by the J_τ. Then $\sigma \in L_s$ if and only if $\sigma|k \le s$ under the lexicographic ordering on the set of sequences of length k and hence the islet I_s is equal to $L_s \setminus \bigcup_r L_r$, where r runs through the set of sequences of length k which are strictly less than s. Finally we know that the islets generate $\mathcal{B}(\sum)$.

THEOREM. Let E and Ω be two metrizable spaces [1], π be the projection of $E \times \Omega$ onto Ω, A be a Souslin subset of $E \times \Omega$ and $B = \pi(A)$. Let $\mathcal{S}$ be the σ-field on Ω generated by the Souslin subsets of Ω. There then exists a mapping h of Ω into E, measurable from σ-fields $\mathcal{S}$ to $\mathcal{B}(E)$ and such that, for all $x \in B$, $(h(x),x) \in A$ (h is a complete cross-section of A).

Proof : Since B is the image of A under a continuous mapping, it is Souslin in Ω, so that $B \in \mathcal{S}$ and the problem reduces to the case where $B = \Omega$. By 78 there exists a continuous mapping f of $\sum$ into $E \times \Omega$ such that $f(\sum) = A$. Let $k = \pi \circ f$. For all $\omega \in \Omega$, $k^{-1}(\{\omega\})$ is a non-empty closed subset of $\sum$; we denote its smallest element

(1) As an exercise, the reader may try to replace in this proof "metrizable" by "Hausdorff".

with respect to the lexicographic ordering by $g(\omega)$. It is easily verified that for all $\tau \in \sum$ the set $g^{-1}(J_\tau)$ is equal to $k(J_\tau)$, the image of J_τ under a continuous mapping and hence Souslin. Thus g is an $\mathcal{S}$-measurable mapping of Ω into $\sum$. It only remains to take $h = \pi' \circ f \circ g$, where π' is the projection of $E \times \Omega$ onto E.

REMARKS. (a) By 17 every Souslin subset of Ω is $\mathcal{B}(\Omega)$-analytic and hence universally measurable by 33. Thus $\mathcal{S} \subset \mathcal{B}_u(\Omega)$ and h is universally measurable. By 18, if Ω is Souslin then also $\mathcal{B}(\Omega) \subset \mathcal{S}$.

(b) Even if $E = \Omega = [0,1]$ and A is Borel with projection equal to Ω, the theorem cannot be improved by constructing a Borel cross-section h or a cross-section with a Souslin graph.

44 should be compared to the abstract form of 81 which we now give : the cross-section constructed here does not depend on a measure on $(\Omega,\mathcal{F})$. More generally the method below can be used to reduce many "abstract" situations to "topological" ones.

THEOREM. Let $(\Omega,\mathcal{F})$ be a measurable space and A an element of $\mathcal{B}(\mathbb{R}_+) \times \mathcal{F}$. Let $\hat{\mathcal{F}}$ be the universal completion of $\mathcal{F}$. There exists an $\hat{\mathcal{F}}$-measurable mapping T of Ω into $\bar{\mathbb{R}}_+$ with the following properties 82

(a) $(T(\omega),\omega) \in A$ for all ω such that $T(\omega) < \infty$

(b) the set $\{T < \infty\}$ is the projection of A onto Ω.

Proof : There exists a separable sub-σ-field $\mathcal{F}'$ of $\mathcal{F}$ such that $A \in \mathcal{B}(\mathbb{R}_+) \times \mathcal{F}'$; replacing $\mathcal{F}$ by $\mathcal{F}'$ if necessary, we can assume that $\mathcal{F}$ is separable. Taking quotients if necessary, we may assume that the atoms of $\mathcal{F}$ are the points of Ω. Then by I.11 we can identify Ω to a subset of $\mathbb{R}$ with the σ-field $\mathcal{B}(\mathbb{R})|_\Omega$. Since A belongs to $\mathcal{B}(\mathbb{R}_+) \times \mathcal{F}$, there exists $A' \times \mathcal{B}(\mathbb{R}_+) \times \mathcal{B}(\mathbb{R})$ such that A is the trace of A' on $\mathbb{R}_+ \times \Omega$. Applying the above theorem to the Lusin set A', we can construct a universally measurable cross-section of A' defined on $\mathbb{R}$. Its restriction h to Ω is universally measurable on Ω (II.32,(2) applied to the injection of Ω into $\mathbb{R}$) and we conclude the proof taking $T(\omega) = h(\omega)$ if ω belongs to the projection of A and $T(\omega) = +\infty$ otherwise.

The second separation theorem

Let $(F,\mathcal{F})$ be a paved set. We say that a subset A of F is $\mathcal{F}$-bianalytic if A and A^c are $\mathcal{F}$-analytic. Bianalytic sets constitute a σ-field $\mathcal{BG}(\mathcal{F})$, which usually contains $\mathcal{F}$, and hence also the σ-field it generates. The typical example is that of a separable metrizable topological space F, with the paving $\mathcal{F}$ of its closed sets ; then the complement of any closed set belongs to $\mathcal{F}_\sigma$, so that $\mathcal{F} \subset \mathcal{BG}(\mathcal{F})$ and the bianalytic σ-field contains the Borel σ-field. 83

We are going to state two equivalent forms of the second separation theorem, and then to give some comments. First, the form due to Novikov.

THEOREM. Assume that $\mathfrak{F} \subset \mathcal{BG}(\mathfrak{F})$. Let (A_n) be a sequence of $\mathfrak{F}$-analytic sets such that $\bigcap_n A_n = \emptyset$. There then exists a sequence (B_n) of $\mathfrak{F}$-bianalytic sets such that $A_n \subset B_n$ for every n and $\bigcap_n B_n = \emptyset$.

The second form is the so called reduction theorem of Kuratowski. To understand it, let us recall how useful the following operation is in set theory : given sets U_n with union U, replace the U_n by smaller disjoint sets V_n with the same union (one usually takes $V_1 = U_1$, $V_2 = U_2 \cap U_1^c$, $V_3 = U_3 \cap U_2^c \cap U_3^c$ ...) This may be called a reduction of the sequence (U_n). Kuratowski's theorem says that reduction is possible in the class of complements of $\mathfrak{F}$-analytic sets (to abbreviate : $\mathfrak{F}$-coanalytic sets) - note that the trivial reduction procedure mentioned above would take us out of this class.

THEOREM. Assume that $\mathfrak{F} \subset \mathcal{BG}(\mathfrak{F})$. Let (U_n) be a sequence of $\mathfrak{F}$-coanalytic sets. There then exists a sequence (V_n) of disjoint $\mathfrak{F}$-coanalytic sets such that $V_n \subset U_n$ for every n and $\cup_n V_n = \cup_n U_n$.

The abstract set-up in these statements is fake generality : the interesting case is that of separable metrizable spaces. In the case of a compact metrizable space, one knows from the first separation theorem that the bianalytic sets are exactly the same as the Borel sets, and the theorem of Novikov becomes very striking (consider for instance the case of a decreasing sequence (A_n)). Most applications concern this compact case, and some of them are very simple and beautiful (Dellacherie [2]).

We have mentioned several times that analytic sets and capacities belong to the "first floor" of descriptive set theory, while the second separation theorem is already on the "second floor". Dellacherie made in [2] the conjecture that the compact case of Novikov's theorem (with "Borel" instead of "bianalytic") still belongs to the first floor. Mokobodzki indeed succeeded in giving an "elementary" proof of it (1), which was further simplified by Dellacherie (2). Another "first floor" proof of a result which includes Novikov's, in the compact case, was given by Saint-Pierre (unpublished).

---

(1) G. Mokobodzki. Démonstration élémentaire d'un théorème de Novikov. Séminaire de Probabilités de Strasbourg, vol. X, 1976, p. 539-543.

(2) Same volume, p. 580.

APPENDIX TO CHAPTER IV

We give here two supplementary results on random sets. The first concerns random sets which have (with non-zero probability) non-countable sections. The second on the contrary determines the structure of random sets almost all of whose sections are countable. The proofs given here are new - at least, we think they are - and use only the "elementary" theory of analytic sets contained in Chapter III and its appendix. Different proofs can be found in Dellacherie [1], Chapter VI, depending on a notion which does not appear in this book, that of "dichotomic capacitances".

We use the following notation : $(\Omega,\mathcal{F}^o,\mathbb{P})$ is a probability space and $\mathcal{F}$ is the completed σ-field of $\mathcal{F}^o$ with respect to $\mathbb{P}$. For some results Ω will be given an increasing family $(\mathcal{F}_t)$ of sub-σ-fields of $\mathcal{F}$ satisfying the usual conditions.

Sets with uncountable sections.

We begin with some preliminary results 109

In the appendix to Chapter III (no. 80) we saw that every non-countable Lusin measurable space L is isomorphic to the interval $[0,1]$; it therefore carries a diffuse measure. This can be shown also in a more elementary fashion : note that L is isomorphic to a Polish space P (III.79) and that P contains (cf. III.80(b)) a compact subset homeomorphic to $\{0,1\}^{\mathbb{N}}$, which also carries diffuse measures (the measure of the "heads or tails game" for example) ; the same result for L then follows. On the other hand we have the following lemma.

LEMMA. *Let $\mathcal{P}$ be the set of probability laws on $\mathbb{R}_+$ with the strict convergence topology. Then the set $\mathcal{D}$ of diffuse laws is a Borel set of $\mathcal{P}$.* 110

Proof : Let $\mu \in \mathcal{P}$ and let $F_\mu(x) = \mu(]-\infty,x])$ be its distribution function. The property that $\mu \in \mathcal{D}$ can be expressed as follows :

for all $n \in \mathbb{N}$ there exists $m \in \mathbb{N}$ such that, for every pair (r,s) of rationals of $[0, n]$ such that $r \leq s \leq r + \frac{1}{m}$, we have $\mu([r,s]) < 1/n$.

This condition is necessary (if μ is diffuse, F_μ is uniformly continuous on $[0,n]$) and sufficient (let $x \in \mathbb{R}_+$; taking $r \leq x$ and $s \geq x$ we see that $\mu\{x\} = 0$). We conclude the proof by noting that $\mu \mapsto \mu([r,s])$ is a Borel function on $\mathcal{P}$.

THEOREM. *Let H be a random set* (1) *belonging to $\mathcal{B}(\mathbb{R}_+) \times \mathcal{F}^o$.* 111

(a) *The set* $c(H) = \{\omega : H(\omega)$ *is non-countable*$\}$ *is $\mathcal{F}^o$-analytic and in particular belongs to* $\mathcal{F}$ (2).

---

(1) As usual, we write $H(\omega) = \{t : (t,\omega) \in H\}$.

(2) This is a special case of the Mazurkiewicz-Sierpinski theorem which states more generally that $c(H)$ is $\mathcal{F}^o$-analytic if H is $(\mathcal{B}(\mathbb{R}_+) \times \mathcal{F}^o)$-analytic.

(b) <u>There exists an</u> $\mathcal{F}^0$<u>-measurable mapping</u> $\mu \mapsto \mu_\omega$ <u>of</u> Ω <u>into</u> $\mathcal{D} \cup \{0\}$ <u>with the following properties</u>

- <u>for all</u> ω, μ_ω <u>is carried by</u> $H(\omega)$.
- $\mathbb{P}\{\omega : \mu_\omega \neq 0\} = \mathbb{P}(c(H))$.

<u>Proof</u> : An immediate argument by monotone classes, starting from the case where H is a rectangle, shows that the mapping $(\mu,\omega) \mapsto \mu(H(\omega))$ on $\mathcal{P} \times \Omega$ is measurable. The set $L = \{(\mu,\omega) : \mu \in \mathcal{D}$ and $\mu(H(\omega)) = 1\}$ is therefore a measurable subset of $\mathcal{D} \times \Omega$. On the other hand, $\mathcal{P}$ is Polish by III.60 and $\mathcal{D}$ is Lusin by 110 : by III.20 (or III.80) the Lusin measurable space $\mathcal{D}$ can be identified with a Borel set of $\mathbb{R}_+$ and hence L can be identified with a measurable subset of $\mathbb{R}_+ \times \Omega$. Finally, by 109, $H(\omega)$ is uncountable if and only if $H(\omega)$ carries a diffuse law, i.e. if ω belongs to the projection of L. Part (a) then follows from III.13 and part (b) from the cross-section theorem III.44-45 - with the zero measure "0" playing here the role of a point adjoined to $\mathcal{D}$.

112 REMARKS. (1) Using the cross-section theorem III.82, instead of III.45, we may construct a mapping $\omega \mapsto \mu_\omega$ of Ω into $\mathcal{D} \cup \{0\}$ which is measurable with respect to the universal completion σ-field $\hat{\mathcal{F}}$ of $\mathcal{F}^0$ and such that $\{\omega : \mu_\omega \neq 0\}$ is equal to $c(H)$.

(2) The function

$$C_H = \inf\{t : H(\omega) \cap [0,t] \text{ is non-countable}\}$$

on Ω is called the <u>penetration time</u> into H. It follows easily from the above theorem that C_H is $\mathcal{F}$-measurable - and is a stopping time if H is a progressive set.

(3) We define a topology on $\mathbb{R}_+$ - which we call the <u>condensation topology</u> in these few numbers - as follows : a point x is considered a cluster point of $E \subset \mathbb{R}_+$ if $x \in E$ or if x is a condensation point of E, i.e. if every ordinary neighbourhood of x in $\mathbb{R}_+$ contains a non-countable infinity of points of E. The corresponding notions of limits will here be denoted by symbols of the type $\text{cd-}\limsup_{y \to x} f(y)$ ... and we leave it to the reader (along with the verification of the topology axioms) to decide the meaning of symbols of the type $\text{cd-}\limsup_{y \downarrow x}$ and $\text{cd-}\limsup_{y \to x, y \neq x}$. We are interested here in the fact that for every progressive (resp. measurable) process (X_t) the process

$$\tilde{X}_t(\omega) = \text{cd-}\limsup_{s \to t, s \neq t} X_s(\omega), \tag{112.1}$$

and the analogous processes constructed by changing the type of the limit, are progressive (resp. $\mathcal{B}(\mathbb{R}_+) \times \mathcal{F}$-measurable). More precisely, there are results similar to those of 89 and 90 with an analogous proof : if X is the indicator of a set H, consider the process $(A_t)_{t > 0}$ defined by $A_t(\omega) = \sup\{s < t : s$ is a condensation point of $H(\omega)\}$ and use penetration times instead of debuts.

When (X_t) is the indicator of a random closed set A, $(\tilde{X}_t)$ is the indicator of a random closed set A (relative to the completion σ-field $\mathcal{F}$). For all ω, the section $\widetilde{A}(\omega)$ is the set of condensation points of A(ω), i.e. the largest perfect set contained in A(ω), the so called perfect kernel of A(ω), and it is well known that for all ω the set $A(\omega)\setminus\widetilde{A}(\omega)$ is countable (Cantor-Bendixson theorem). We often say that the random set $\widetilde{A}$ itself is the perfect kernel of A.

(4) We shall see in a later chapter that if H is optional or predictable the family $\omega\mapsto\mu_\omega$ of theorem 111 can be subjected to additional conditions of optionality or predictability.

Theorem 111 can be improved as follows :

THEOREM. Suppose that H is a random closed set. Then it can be assumed that the mapping $\omega\mapsto\mu_\omega$ of assertion (b) has the following property 113

for almost all $\omega\in\Omega$ the support of μ_ω is the perfect kernel $\widetilde{H}(\omega)$ of H(ω).

Proof : We arrange all the pairs of rationals ≥ 0 (r,s) such that r < s in a sequence (r_n,s_n) and for all n let $H_n = H\cap]r_n,s_n[$. We construct a mapping $\omega\mapsto\mu^n_\omega$ satisfying assertion (b) of 111 relative to H_n and set $\lambda_\omega = \sum_n 2^{-n}\mu^n_\omega$. For every ω this is a diffuse positive measure of mass ≤ 1 carried by H(ω) and hence also by the perfect kernel $\widetilde{H}(\omega)$ and, for almost all ω, λ_ω is different from zero on every open interval I such that $I\cap H(\omega)$ is non-countable, i.e. every open interval intersecting $\widetilde{H}(\omega)$. The support of λ_ω is therefore exactly $\widetilde{H}(\omega)$. To construct the required mapping it only remains to set $\mu_\omega = \lambda_\omega/\|\lambda_\omega\|$ of $\lambda_\omega\neq 0$ and $\mu_\omega = 0$ if $\lambda_\omega = 0$.

REMARK. As in no. 112, (1), it is also possible to construct a mapping $\omega\mapsto\mu_\omega$ which is measurable with respect to $\hat{\mathcal{F}}$(but not to $\mathcal{F}^0$), such that the support of μ_ω is $\widetilde{H}(\omega)$ for all ω.

Derived sets of random sets

We saw that if H is a random closed set, $H\setminus\widetilde{H}$ is a random set with countable sections. We shall show in these few numbers that it is indistinguishable from a countable union of graphs. More generally, since the sections of $H\setminus\widetilde{H}$ are $\mathcal{G}_\delta$ of $\mathbb{R}_+$, we shall establish this result for all "random $\mathcal{G}_\delta$" with countable sections.

We recall first a classical definition and give a simple lemma on random sets.

DEFINITION. Let F be a subset of $\mathbb{R}_+$. The derived set [1] of F, denoted by F', is the set of points of F which are not isolated in F. For every countable ordinal α the derived set of F of order α is defined by transfinite induction as follows 114

(114.1)
$$F^0 = F \ ; \ F^{\alpha+1} = (F^\alpha)'$$
$$F^\beta = \bigcap_{\alpha<\beta} F^\alpha \text{ for every limit ordinal } \beta.$$

---

(1) To recover exactly the classical notion we should say "derived set of F in F".

If H *is a subset of* $\mathbb{R}_+ \times \Omega$ *we denote by* H', H^α *the subset of* $\mathbb{R}_+ \times \Omega$ *whose section for all* $\omega \in \Omega$ *is* $H(\omega)'$, $H(\omega)^\alpha$.

115 THEOREM. *If* H *belongs to* $\mathcal{B}(\mathbb{R}_+) \times \mathcal{F}$ *(resp. if* H *is a progressive set), for every countable ordinal* α, H^α *is a* $(\mathcal{B}(\mathbb{R}_+) \times \mathcal{F})$*-measurable (resp. progressive) random set and* $H \backslash H^\alpha$ *is a countable union of* $\mathcal{F}$*-measurable functions (resp. stopping times).*

Proof : It is sufficient to argue with H'. For simplicity we suppose that H is progressive. For every rational r set $T^r = \inf\{t > r : t \geq 0$ and $(t,\omega) \in H\}$; the isolated points of $H(\omega)$ appear among the points $T^r(\omega)$; and $T^r(\omega)$ is isolated if and only if

$$(T^r(\omega),\omega) \in H(\omega) \ ; \ r < T^r(\omega) \ ; \text{ there exists } \varepsilon > 0$$
$$\text{such that } H(\omega) \cap [T^r(\omega), T^r(\omega) + \varepsilon] = \emptyset.$$

Let A(r) be the set of ω with these properties ; it is not difficult to verify that A(r) belongs to $\mathcal{F}_{T^r}$ and H\H' then is the union of the graphs of the stopping times $T^r_{A(r)}$ (53). This implies that H' is progressive.

In the following statelement we omit the case of progressive sets.

116 THEOREM. (a) *Let* H *be a* $(\mathcal{B}(\mathbb{R}_+) \times \mathcal{F})$*-measurable set. There exists a countable ordinal* α *such that* H^α *and* $H^{\alpha+1}$ *are indistinguishable.*

(b) *Suppose that the sections of* H *are* $\mathcal{G}_\delta$*'s. Then the sections* $H^\alpha(\delta)$ *are a.s. (for the above ordinal* α *) either empty or non-countable. In particular, if the section of* H *are countable* $\mathcal{G}_\delta$, $H^\alpha(\omega)$ *is empty for almost all* ω *and* H *is indistinguishable from a countable union of graphs of* $\mathcal{F}$*-measurable functions.*

Proof : For every rational r let $T^\alpha_r(\omega) = \inf\{t > r : (t,\omega) \in H^\alpha\}$. Since H^α decreases as α increases, T^α_r increases and no. 0.8 applied to the numbers $E[\exp(-T_r)]$ shows that there exists a countable ordinal α such that $T^\alpha_r = T^{\alpha+1}_r$ a.s. Since there are only countable many rationals, there exists an α such that this is true simultaneously for all rationals. Then we consider an ω such that $T^\alpha_r(\omega) = T^{\alpha+1}_r(\omega)$ for all rational r ; $H^\alpha(\omega)$ has no isolated point - for if s were such a point, there would also exist an interval]u,v[with rational end-points containing s and not intersecting $H^{\alpha+1}(\omega)$, and we would have $T^\alpha_u(\omega) \leq s$, $T^{\alpha+1}_u(\omega) \geq v$ contrary to the hypothesis. Hence $H^\alpha(\omega) = H^{\alpha+1}(\omega)$.

Suppose that the sections $H(\omega)$ are $\mathcal{G}_\delta$'s. We saw earlier that $H(\omega)\backslash H^\alpha(\omega)$ is a countable set D and hence $H^\alpha(\omega) = H(\omega) \cap D^c$ is a $\mathcal{G}_\delta$. We choose ω such that $H^\alpha(\omega)$ is equal to $H^{\alpha+1}(\omega)$, i.e. has no isolated points and is non-empty ; then $H^\alpha(\omega)$ is a non-empty $\mathcal{G}_\delta$ of $\mathbb{R}_+$ and hence a Baire space and it cannot be countable (since every countable Baire space has at least one isolated point).

If the sections $H(\omega)$ are countable $\mathcal{G}_\delta$, $H^\alpha(\omega)$ is a.s. empty and hence H is indistinguishable from $H\backslash H^\alpha$, which is a countable union of graphs by 115.

REMARKS. (1) In the last assertion the $\mathcal{F}$-measurable random variables might of course be replaced by $\mathcal{F}^{o}$-measurable random variables.

(2) This theorem applies in particular to sets with closed sections. In this case there exists an ordinal α such that H^{α} is indistinguishable from the perfect kernel $\widetilde{H}$ of H and $H \setminus \widetilde{H}$ is indistinguishable from a countable union of graphs.

Sets with countable sections

THEOREM. *Let* H *be a* $\mathcal{B}(\mathbb{R}_+) \times \mathcal{F}$*-measurable (resp. optional, predictable) set with countable sections*. *Then* H *is indistinguishable from a countable union of graphs of* $\mathcal{F}$*-measurable functions* (*resp. optional, predictable times*). 117

Proof : We can pass to the optional and predictable cases using no. IV.88 (and this only requires that the filtration satisfy the usual conditions). We can therefore forget about the filtration. The theorem follows from no. 116 for sets whose sections are $\mathcal{G}'^{s}_{\delta}$ and from the following lemma (which requires only that the σ-field $\mathcal{F}$ be complete).

LEMMA. *Let* H *be a* $(\mathcal{B}(\mathbb{R}_+) \times \mathcal{F})$*-measurable set*. *Then there exist a* $(\mathcal{B}(\mathbb{R}_+) \times \mathcal{F})$*-measurable set* K, *all of whose sections are* $\mathcal{G}'^{s}_{\delta}$, *and a Borel mapping* h *of* $\mathbb{R}_+$ *into* $\mathbb{R}_+$ *such that* $(t,\omega) \mapsto (h(t),\omega)$ *induces a bijection of* K *onto* H.

The lemma is applied as follows : by 116, K is indistinguishable from the union of the graphs of a sequence of r.v. T_n, and H then is indistinguishable from the union of the graphs of r.v. $h \circ T_n$.

Proof of the lemma : Since there exists a separable sub-σ-field $\mathcal{G}$ of $\mathcal{F}$ such that H belongs to $\mathcal{B}(\mathbb{R}_+) \times \mathcal{G}$, there is no loss of generality in assuming that $\mathcal{F}$ is separable. Nor is there any such loss (taking if necessary a quotient relative to the equivalence relation associated with $\mathcal{F}$) in assuming that the atoms of $\mathcal{F}$ are the points of Ω. Then $(\Omega,\mathcal{F})$ can be identified (I.11) with a subset of $[0,1]$ given its Borel σ-field. Finally, since every Borel set of $\Omega \times \mathbb{R}_+$ is the trace of a Borel set of $[0,1] \times \mathbb{R}_+$, we can reduce to the case where $\Omega = [0,1]$, $\mathcal{F} = \mathcal{B}([0,1])$ and prove the lemma under these conditions.

The set H is Lusin and hence there exists (III.79, appendix) a *closed* set F of $\Sigma = \mathbb{N}^{\mathbb{N}}$ and a continuous bijection of F onto H, which we denote by $\sigma \mapsto (h_1(\sigma), h_2(\sigma))$, where h_1 and h_2 take values respectively in $\mathbb{R}_+$ and in Ω. We now use the fact [(1)] that Σ is homeomorphic to the set I of irrational numbers of $[0,1]$ to identify F with a closed subset of $I \subset \mathbb{R}_+$ and set

$$K = \{(t,\omega) \in \mathbb{R}_+ \times \Omega : t \in F,\ h_2(t) = \omega\} \quad \text{(K is the graph of } h_2\text{)}.$$

Let h be a Borel extension to $\mathbb{R}_+$ of the mapping h_1 defined on F. The mapping

(1) See the remark below.

$(t,\omega) \mapsto (h(t),\omega)$ then is Borel and induces a bijection of K onto H. On the other hand, for all ω the section $K(\omega)$ is closed in F since h_2 is continuous, hence closed in I, and since I is a $\mathcal{G}_\delta$ of $\mathbb{R}_+$, so is $K(\omega)$.

118 REMARK. The traditional homeomorphism between I and $\Sigma = \mathbb{N}^{\mathbb{N}}$ is given by the contined fraction expansion of irrational numbers. We now give a means of avoiding it. With every sequence $\sigma = (n_1, n_2, n_3, \ldots) \in \Sigma$ we associate the number $f(\sigma) \in [0,\frac{1}{2}]$ whose dyadic expansion begins with n_1 "0"s and continues with n_2 "1"s, then n_3 "0"s, n_4 "1"s etc .... Since all the n_i are > 0, the function f thus defined is a bijection between Σ and the complement J of the dyadic rationals in $[0,\frac{1}{2}]$, and it is easily verified that f is a homeomorphism of Σ onto J. Since J is a $\mathcal{G}_\delta$ of $[0,\frac{1}{2}]$; the reader may just replace I by J in the above proof.

COMMENTS

In this chapter on notation we have added a small paragraph on transfinite induction along countable ordinals, a procedure which we find extremely pleasant and intuitive, but which doesn't belong to the standard background of most students in probability (at least in France). Many people think that transfinite induction is "superseded" by such general results as Zorn's lemma. This isn't true : Zorn's lemma may spare much work with well ordered sets in set theory or topology, but doesn't replace naturally the typically "countable" transfinite inductions of real analysis.

As far as the continuum hypothesis is concerned, it seems that its consequences for real analysis (Mokobodzki's theorems on rapid filters and medial limits, Mokobodzki's lifting theorem) can also be deduced from more complicated, but "softer", axioms of set theory ("Martin's axiom"), which are also compatible with the negation of the continuum hypothesis.

CHAPTER I

The contents of this chapter are classical. Compared to the first edition, we have entirely omitted non-metrizable compact spaces, and added the fact, classical but not too well known, that Hausdorff separable measurable spaces can be inbedded into [0,1]. We have also given more flexible variants of the monotone class theorem.

CHAPTER II

This chapter also is quite classical. The results given without a proof belong to any text on measure theory. Compared to the first edition, we have omitted a small section on Radon measures (the subject is taken over in chapter III in a modified form) and added the Vitali-Hahn-Sakes theorem, the converse of the Dunford-Pettis compactness criterion (i.e. the implication that weak compactness $\Rightarrow$ uniform integrability) and Mokobodzki's theorem on rapid filters (to be completed later on by his theorem on medial limits). Finally, we have introduced generalized conditional expectations in a slightly non classical way (39, b)). All this has been added, not just for beauty's sake, but because it has proved useful either in martingale theory or in the theory of Markov processes.

CHAPTER III

For probabilists, the main use of analytic sets and capacities is the measurability theorem for debuts (44). On this subject, the first sentence of the chapter which ascribes to Hunt the discovery of the relation between this measurability problem and capacity theory, isn't quite true. Doob's first fundamental paper on Brownian motion (Doob [4]), which partially inspired Hunt's great work, contained a proof of measurability of a debut - the hitting time of a Borel set - deduced, not from capacity theory which didn't exist yet, but from results of Cartan in classical potential theory. Moreover both Hunt and Doob proved more than the measurability of a debut, namely the possibility of approximating hitting times of Borel sets by hitting times of open sets. In the general case this uses the full strength of Choquet's theory, whereas Blackwell remarked (1) that the measurability theorem for debuts just requires that analytic sets should be universally measurable, a result known long before Choquet (Saks [1]).

Similarly, the cross section theorems of chapter IV depend only on 44 b), which is also established in the appendix without using capacities. The reader might therefore be tempted to believe that one could do without Choquet's theorem. This is wrong : Choquet's theorem is at the heart of measure theory, and it is much better to learn it rather than the tricks that may replace it in each particular case.

This volume contains much more on analytic sets than the first edition. Blackwell's theorem is given a more prominent role, as a result which is at the same time extremely simple and powerful, a remarkable tool for discovery (even if more "elementary" proofs come to light afterwards). We have systematically introduced Lusin, etc, measurable spaces, partly under the influence of Yen [1]. Finally, and perhaps above all, we hope that our simple proof will dissipate the fear of the magnificent Souslin-Lusin theorem, so often considered as "deep" (with the connotation of "slippery and useless"), because of the complication of its usual proofs (Bourbaki [3], for example).

The first edition contained some numbers devoted to "regular measures" in an abstract setting. The only remnant of this is the notion of a semi-compact paving (which is defined, but practically used nowhere). The reason for this omission is the importance given to Radon measures on non locally compact spaces, a theory created by Prokhorov in [1], developed for probabilistic (even statistical) purposes by Le Cam, Varadarajan, Parthasarathy, in relation with much research on measure theory in topological linear spaces (inspired by an idea of Gelfand [1] and carried on by Minlos and the Russian school, by Schwartz, Fernique...). This theory seems to have reached maturity, and we have presented it following Bourbaki [5], with more

(1) In an unpublished report, which was the first paper to investigate the measurability of debuts for arbitrary measurable processes, and was thus the starting point of the "general theory of processes". The only copy we had of it has been lost.

emphasis on capacity methods. It might interest the reader to compare our proof of the regularity of measures on a Polish space to Prokhorov's beautiful original proof in [1].

This doesn't mean, however, that abstract regular measures won't come again to the foreground. A good reference may be the small lecture notes set by Pfanzagl-Pierlo [1].

On "standard" spaces, i.e. Lusin spaces as defined by Bourbaki, and on their applications to random distributions, Fernique [1] is a good reference. As we already said in the main text, light on this subject came from a remark of Cartier [1].

The theorem on bimeasures is mentioned, without proof, in Kingman [1]. Cartier mentioned to us that it could probably be proved along the same lines as C. Doléans-Dade's proof of the supermartingale decomposition theorem, and this was successfully carried out by Morando [1]. The theorem has been rediscovered several times since, and turned out to have nice applications.

We now give some additional references. The renewal of interest in analytic set theory (1) has led to several new books on the subject, among which Hoffmann-Jørgensen [1] and Christensen [1] are highly commendable. In particular, the latter contains a study of the so called Effros structure, a measurable structure on the family of all closed subsets of a Polish space, which has important applications. These two books like the present work, deal only with the elementary levels of analytic set theory (see the appendix to chapter III). To climb to the level just above, a possible guide is Dellacherie [2] (partly following Rogers [1]), or Dellacherie-Meyer [1] in a language more familiar to probabilists. Logicians have climbed far higher summits in descriptive set theory, but we are unable to follow them : a book by Moschovakis should be forthcoming on this subject.

As far as capacities are concerned, the reader may consult the memoir [1] of Choquet, the spring from which the whole theory took its rise, still full of lively ideas. Also, the excellent papers by Sion, to whom we owe much. Finally, papers giving applications of capacities to classical analysis : Brelot [1], [2], Helms [1] and Carleson [1] (and the forthcoming books on classical potential theory in its relation to Brownian motion, bu Doob and Rao (2)). For the happy few that have access to them, we strongly commend the collection of seminars of the University of Paris : Séminaire Brelot-Choquet-Deny (potential theory) and Séminaire Choquet (initiation to analysis) : some of them appear now in Springer's Lecture Notes series, but the intermediate volumes are difficult to find.

(1) This renewal has also led to more general definitions of analytic sets in non metrizable non separable topological spaces. This scarcely concerns probabilists, and we give no indication concerning this direction of research, in spite of its intrinsic interest.

(2) Rao's little book has just appeared : Brownian motion and classical potential theory, Aarhus Lecture Notes series no. 47, Aarhus Univ. 1977.

CHAPTER IV

This chapter has been deeply changed since the first edition. We have reduced the role of separability (omitting in particular a ridiculous appendix on "abstract" separability and all that concerned the "second canonical process"), and added a number of sections on subjects which, from our experience of several years, we now consider important : the measurability properties on canonical spaces, essential limits, and so on. On the other hand, we have included in the new edition of chapter IV many results on the "general theory of processes" which were spread over chapters VII and VIII of the first edition.

The "general" theory of processes is often criticized as too abstract, but it is hard to know what is meant here, since after all mathematics as a whole is an abstract science. If "abstract" means "difficult to learn", or even "boring", we may agree to a certain extent, but not if it is meant that it is a mere "divertissement dans le goût français". Indeed, the general theory of processes hasn't grown up as an autonomous branch of mathematics, but rather in constant interplay with martingale theory and the theory of Markov processes, as the set of tools necessary to fulfill the program of Lévy, Doob, Ito, Chung, Hunt, and in favour of which Chung has for years conducted an unwearying propaganda : "Look at the sample functions...". Some examples : Increasing families of σ-fields and stopping times come from martingale theory (Doob, around 1940), and even from gambling theory, the very birthplace of all probability. The idea of systematically extending to stopping times results known to hold at fixed times is an imitation of the "strong Markov property", first mentioned in a paper of Doob's on Markov chains in 1942, then by Hunt for processes with independent increments, then independently for general Markov processes by Blumenthal and the school of Dynkin in the USSR. We also owe to the latter many useful results on progressive processes and stopping times. The idea of a process "with a lifetime", before it became a general tool under the shape of killing operators, occurred in concrete cases of "explosive" Markov chains or diffusions (e.g. in population problems), while the little trick of adding a fictitious state is due to Doob [2]. The idea of a predictable or foretellable stopping time appears (with a wrong proof) in Meyer's thesis in 1962, in connection with the theory of Markov processes, while that of a totally inacessible stopping time is implicit in the quasi-left-continuity of Markov processes, a very important property discovered by Hunt and Blumenthal (and the most intuitive example of stopping times, namely the successive jumps of a Poisson process, turn out to be totally inaccessible). The optional σ-field and cross-section theorem appear in a seminar report of 1963 (Meyer [4]) with a wrong proof (1).

(1) The first simple (and correct) proof of the cross-section theorems was discovered by Cornea-Licea, then further simplified by Dellacherie (see Meyer [3]).

Interest in the predictable σ-field grew from the work of Catherine Doléans : [1], on a new proof of the supermartingale decomposition theorem along lines suggested by Cartier, and [2] on stochastic calculus, a branch which is by essence concrete, since it has applications even to engineering. Finally, σ-fields of type $_{T-}$ had been considered by Chung and Doob [1], and their relation to predictability was discovered while Chung was at Strasbourg (1967-68), with motivations coming from the theory of Markov chains (much of the theory took its definitive form under pressure from Chung).

The use of up- and downcrossings is borrowed from martingale theory (Doob [1], p. 315-316). Doob [5] (on the heat equation) was our model for the transfinite induction arguments on intervals where functions oscillate little. The use of essential topology on the line comes from time reversal for Markov processes (Chung-Walsh [1], Doob [6], Walsh [1]; for a particularly nice application, see Walsh [2]).

The point of view that consists in rejecting instantaneous observations on processes, to be replaced by averages on small time intervals, has its source in the work on random distributions. The method we present goes back to an old discussion with Cartier. Ito studied (and explained once in a visit to the Strasbourg seminar) a similar point of view ([1], [2]) aiming more than we do to the study of sample path regularity. Some recent and important work of Knight [1] pushes the theory further.

Our paragraph 3 also differs from the earlier accounts of the theory (the first edition of this book, Dellacherie [1], Meyer [5]) by an important feature : the partial rejection of the "usual conditions", at the cost of some technical complication. Let us say again that we haven't done it just for the sake of developing a more general theory, but because it turns out to be useful (for instance, in the work of Azéma on time reversal, of Foellmer on quasimartingales, etc.).

The theory of processes given in this volume still is incomplete : it still lacks the "projection theorems", which need martingale theory and will be given later, and the criteria for right and left continuity of optional or predictable processes (Mertens [1], Dellacherie [1], p.101).

INDEX OF TERMINOLOGY

Accessible stopping time, IV.80.
Adapted process, IV.12.
Additive set function, III.30.
Additivity, countable, II.1.
Algebra, Boolean, I.1.
σ-algebra, I.1, note.
Almost complete cross-section, III.44.
Almost-equivalent processes, IV.35.
Almost-modification, IV.35.
Almost sure, II.2, convergence, II.10.
Analytic set in a separable metrizable space, III.15.
$\mathfrak{F}$-analytic function, III.61.
$\mathfrak{F}$-analytic set, III.7.
Approximation, Lebesgue, I.17.
Augmentation, usual, IV.48.

Baire σ-field, I.7.
Base space of a process, IV.1.
Bianalytic set, III.83 (Appendix).
Bimeasure, III.74.
Blackwell space, III.24.
Blackwell's theorem, III.26.
Borel σ-field, function, set, I.7.
Borel isomorphism, I.11.

Càdlàg process, IV.16, footnote.
Canonical process, IV.9, IV.94, etc.
Capacitable, III.27.
Capacitance, III.29.
Carathéodory's extension theorem, III.34.
Choquet's theorem, III.28.
Closure of a random set, IV.31, of a family of subsets, 0.2.
Coanalytic set, III.16, footnote 2).
Compact paving, III.3.
Complete cross-section, III.44, footnote 3).
Complete filtration, IV.48.
Complete probability space, II.3 and II.32.
Completion, II.32, universal, II.32.
Conditional expectation, II.37 and II.40, generalized, II.39.

Conditional independence, II.43.
Conditional probability, II.38.
Conditions, usual, IV.48.
Continuum hypothesis, 0.8.
Convergence, dominated, II.6, of r.v., II.10, narrow, III.54.
Cosouslin measurable or metrizable space, III.16.
Cross-section theorems, III.44-45, III.81 (appendix), IV.84 (optional), IV.85 predictable).

Daniell's theorem, III.35.
Debut of a subset of $\mathbb{R}_+ \times \Omega$, III.44, essential, IV.39.
Degenerate law, II.4.
Derived set of a random set, IV.114 (Appendix).
Determining system, III.75 (Appendix).
Deterministic filtration, IV.13.
Dichotomic capacitance, IV. Appendix.
Diffuse law, II.19, note.
Disintegration of measures, III.70-73.
Distribution of a r.v., II.11.
Dominated convergence theorem, II.6.
Downcrossings, number of, IV.21.
Dunford-Pettis weak compactness criterion, II.25.

Elementary r.v., I.13.
Equivalent processes, IV.4.
Essential debut, IV.39.
Essential least upper bound, II.8.
Essential limit, topology, IV.36.
Evanescent random set, IV.8.
Event prior to t, IV.11, prior to a stopping time, IV.52, strictly prior to a stopping time, IV.54.
Expectation, II.5, conditional, II.37-40, generalized conditional, II.39.

Fatou's Lemma, II.7.
σ-field, I.1, Baire, Borel, I.7, generated by..., I.5, optional, predictable, IV.61, product, I.8, progressive, IV.31.
Filter, rapid, II.27.
Filtration, IV.11, complete, IV.48, deterministic, IV.13, left-quasi-continuous, IV.82, natural, IV.12.
Finite intersection property, III.2.
Fubini's theorem, II.14.

Function, analytic, II.61, Borel, 1.7, free of oscillaroty discontinuities, IV.20, measurable, I.2, universally measurable, II.32.

Galmarino's test, IV.99-101.
Generalized conditional expectation, II.39.
Generated by..., σ-field, I.5.

Hypothesis, continuum, 0.8.

Image law, II.11.
Increasing family of σ-fields, IV.11, cf. also Filtration.
Independence, II.33-34, conditional, II.43.
Indicator, I.3.
Indistinguishable processes, IV.7.
Induction, transfinite, 0.8.
Inner regular measure, III.46.
Integrable, uniformly, set, II.17.
Integral of a family of laws, II.15.
Internally negligible set, II.30.
Interval, stochastic, IV.60.
Inverse system, III.53.
Isomorphism, measurable or Borel, I.11.

Jensen's Inequality, II.41. (4).

Kernel, of a Souslin scheme, III.75 (appendix), perfect, IV.112 (appendix).
Kind, ordinal of the first or second, 0.8.
Kolmogorov's theorem, II.51-52.

(L), (LL), (LLL) space, III.63.
La Vallée Poussin's uniform integrability criterion, II.22.
Law (probability), degenerate, II.4, image, II.11, of a r.v., II.11, product, II.14 (Remark), time, of a process, IV.4.
LCC space, 0.7.
Lebesgue approximation, I.17.
Lebesgue measurable process, IV.35.
Lebesgue's theorem, II.6-7.
Left-continuous capacity, II.39.
Lemma, Fatou's, II.7.
Lifetime, IV.19.
Limit, inverse, III.53.
Limit ordinal, 0.8.
Lindelöf space, III.63.
Locally bounded measure, III.47.
L.s.c. functions, 0.7.

Lusin topological space, III.67.
Lusin measurable space of metrizable topological space, III.16.

Mapping, see Function.
Mass, unit, III.4.
Measurable family of laws, II.13.
Measurable function, I.2.
Measurable process, IV.3, in the Lebesgue sense, IV.35, progessively, IV.14.
Measurable space, I.1.
Measure, 0.5, convergence in, II.10, inner regular, III.46, locally bounded, III.47, outer regular, III.47, product, II.14, Radon, III.46.
Monotone class, I.19, theorem, I.19-21.

Narrow convergence of measures, III.54.
Natural family of a process, IV.12.
Negligible, internally, set, II.30.
Number of upcrossings and downcrossings, IV.21.

Operation (A), Souslin, III.75 (Appendix).
Optional σ-field, process, IV.61.
Optional time, IV.49.
Ordinal, 0.8., limit, of the first pr second kind, 0.8.
Orthogonality, II.9.
Outer regular measure, III.47.

Path, IV.1.
Paved set, III.1.
Paving, III.1, compact, semicompact, III.3, sum, product, III.1.
Penetration time, IV.111 (Appendix).
Perfect random set, IV.112 (Appendix).
Polish space, 0.7.
Predecessor of an ordinal, 0.8.
Predictable σ-field, process, IV.61.
Predictable time, IV.68.
Prior to t, event, IV.11.
Prior to T, event, IV.52.
Probability, convergence in, II.10.
Probability law, II.1.
Probability space, II.1, complete, II.3.
Process, IV.1, adapted, IV.12, canonical, IV.9, measurable, IV.3, measurable in the Lebesgue sense, IV.35, progressive, IV.14, separable, IV.26.
Processes, equivalent, IV.4, indistinguishable, IV.6.
Product σ-field, I.8.
Product law measure, II.14.

Product paving, III.1.
Progressive, progressively, measurable, IV.14.
Preogressive σ-field, set, IV.31.
Prokhorov's Theorems, (inverse limits), III.53, (strict compactness), III.59.
Pseudo-law of a process, IV.44.
Pseudo-path, IV.41.

Quasi-left continuous filtration, IV.82.

Radon measure, III.46.
Random closed (right-, left-closed) set, IV.31.
Random set, IV.8, measurable, progressive, closed..., IV.31.
Random variable, r.v., I.2, elementary, I.13, real-valued, I.13.
Rapid filter, II.27.
R.c.l.l., etc., IV.16.
Real-valued random variable, I.13.
Regular, inner, measure, III.46.
Regular, outer, measure, III.47.
Regular Souslin scheme, III.75 (Appendix).
Riesz Representation Theorem, III.35 bis.
Right-continuous capacity, III.41.
Right-continuous filtration, IV.11.
Right essential limit, topology, IV.36.
Right topology, IV.24, note.

Scheme, Souslin, III.75 (appendix).
Semicompact paving, III.3.
Separable σ-field, I.10.
Separable measurable space, I.10.
Separable, right separable process, IV.25-26.
Separable topological space, 0.7.
Separable theorem, III.14 and III.23, second, III. Appendix.
Set, analytic, in a metrizable space, III.15, $\mathfrak{F}$-analytic, III.7, $\mathfrak{F}$-bianalytic, III.83 (appendix), capacitable, III.27, internally negligible, II.30, paved, III.1, time, of a process, IV.1, universally measurable, II.32.
Sets separable by a paving, III.14.
Skorokhod topology, IV.19.
Souslin, III.16 and III.67.
Souslin scheme, III.75 (appendix).
Souslin-Lusin theorem, III.21.
Space, Blackwell, III.24, Hausdorff, I.9, Lusin, Souslin, cosouslin, III.16, measurable I.1, separable, I.10.
Standard modification of a process, IV.6.
State of a process at instant t, IV.1.

State space of a process, IV.1.
Stationary sequence, IV.55.
Stochastic interval, IV.60.
Stopping time, IV.49, accessible, totally inaccessible, IV.80, foretellable, IV.70, predictable, IV.68, in the wide sense, IV.49.
Strictly prior to, event, IV.54.
Strong convergence, II.10.
Strongly subadditive set function, III.30.
Sum paving, III.1.
Support of a measure, III.50.
System, determining, III.75 (Appendix), inverse, of laws, III.53.

Test, Galmarino's, IV.99-101.
Time, accessible, totally inaccessible, IV.80, penetration, IV.111 (appendix), predictable, IV.68, optional, IV.49.
Time law of a process, IV.4.
Time set of a process, IV.1.
Topological space, LCC, 0.7, Lindelöf ((L), (LL), (LLL)), III.63, Lusin, Souslin, cosouslin, III.16 (metrizable case) and III.67, Polish 0.7, separable, 0.7.
Topology, essential, right essential, IV.36.
Totally inaccessible, IV.80.
Transfinite induction, 0.8.

Uniformly integrable, II.17.
Unit mass, II.4.
Universal completion, II.32.
Universally measurable, II.32.
Upcrossings, number of, IV.21.
U.s.c. functions, 0.7.
Usual augmentation, conditions, IV.48.

Variable, random, I.2.
Versions of a process, IV.4.
Vitali-Hahn-Saks theorem, II.23.

Weak convergence, II.10.
Well measurable (= optinal) process, set, IV.61.
Wide sense stopping time, IV.49.

INDEX OF NOTATION

A^c, A, $A\backslash B$, $A \Delta B$, $\{x : \ldots\}$, $f|_A$, $\mathcal{E}|_A$ (set theory), 0.1.

closure under $\cup f$, $\cup c$, $\cup a$, $\cup mc$,..., 0.2.

$\mathcal{E}_\sigma, \mathcal{E}_\delta, \mathcal{E}_{\sigma\delta}, \ldots$, 0.2.

$f \vee g$, $f \wedge g$, f^+, f^-, $\bigvee_i \mathcal{E}_i$, 0.3.

$s \uparrow t$, $s \uparrow\uparrow t$, $s_n \uparrow t$, $\lim_n$, $\liminf_n$, 0.4.

$\|\mu\|$, total mass of the measure μ, $\int f\mu, \lambda/\mu$, $\int f(t)dF(t)$, 0.5.

$\mathcal{C}(E)$, $\mathcal{C}_b(E)$, $\mathcal{C}_c(E)$, $\mathcal{C}_0(E)$, $\mathcal{C}_c^\infty(E)$ spaces of continuous functions, 0.6.

$\mathfrak{m}(\mathcal{E})$, $b(\mathcal{E})$, measurable functions, 0.6.

l.s.c., u.s.c. (semi-continuity), 0.7.

r.v., random variable, I.2.

I_A, indicator of A, I.3.

$\mathfrak{T}(\mathcal{G})$, σ-field generated by $\mathcal{G}$, 1.5.

$\mathfrak{F}(f_i, i \in I)$, σ-field generated by the f_i, I.5.

$\mathcal{B}(E)$, Borel σ-field, I.7.

$\prod_i \mathcal{E}_i$, $\mathcal{E}_1 \times \mathcal{E}_2$, product σ-field, I.8.

$(\Omega,\mathfrak{F})$, Hausdorff space associated with $(\Omega,\mathfrak{F})$, I.9.

$f \otimes g$, the mapping $(x,y) \longmapsto f(x)g(y)$, I.24 (note).

a.s., almost surely, II.1.

ε_x, unit mass at x, II.4.

$\mathbb{E}[f]$, expectation of f, II.5.

$\mathcal{L}^p$ (function space), L^p (class space), II.8.

$\|f\|_p$, norm in L^p (possibly $+\infty$), II.8.

$\sigma(L^1,L^\infty)$, $\sigma(L^2,L^2)$ weak topologies, II.10.

$f(\mathbb{P})$, image law of $\mathbb{P}$ under f, II.11.

$\lambda \otimes \mu$, product measure, II.14 (Remark (b)).

$\int \mathbb{P}_x \mathbb{Q}(dx)$, integral of probability laws, II.15.

f^c, f_c, truncation of f, II.17.

$\mathfrak{F}^{\mathbb{P}}$, completion σ-field of $\mathfrak{F}$ with respect to $\mathbb{P}$, II.32.

$\hat{\mathfrak{F}}$, universal completion σ-field of $\mathfrak{F}$, II.32.

$\mathcal{B}_u(E)$, universally measurable σ-field, II.32.

$\mathbb{E}[X/f]$, conditional expectation (provisional notation), II.37.

$\mathbb{E}[X|\mathcal{E}]$, $\mathbb{E}[X|f_i, i \in I]$, $\mathbb{P}[A|\mathcal{E}]$, $\mathbb{P}[A|f_i, i \in I]$, conditional expectations and probabilities, II.40.

$\mathbb{E}[X|\mathcal{E}_1|\mathcal{E}_2]$, iterated conditional expectation, II.40.

$\prod_i \mathcal{E}_i$, $\mathcal{E}_1 \times \mathcal{E}_2$, product paving, ambiguous notation used only in Chapter III.

$\sum_i \mathcal{E}_i$, sum paving.

$\mathcal{K}(E)$, paving consisting of the compact subsets of a Hausdorff space, III.3.

$\mathcal{A}(\mathcal{E})$, paving consisting of the $\mathcal{E}$-analytic sets, III.7.

$\mathcal{C}(\mathcal{E})$, closure of $\mathcal{E}$ under $(\cup c, \cap c)$, notation used occasionally in Chapter III, III.14.

$\mathcal{E}|_A$, trace σ-field of $\mathcal{E}$ on A, III.15.

$\mathcal{G}(E)$, paving consisting of the open sets of a topological space.

$\mathcal{A}(E)$, paving consisting of the analytic subsets of a metrizable space E.

$\mathcal{K}$, $\mathcal{G}$, $\mathcal{B}$, $\mathcal{A}$, abridged notation for $\mathcal{K}(E)$, etc...

$\mathcal{L}(\mathcal{E})$, $\mathcal{S}(\mathcal{E})$, $\mathcal{S}'(\mathcal{E})$, pavings consisting of the Lusin, Souslin, cosouslin subsets of a Hausdorff measurable space $(E,\mathcal{E})$.

$\mathcal{A}'(\mathcal{E})$, paving consisting of the complements of elements of $\mathcal{A}(\mathcal{E})$, III.19.

I^*, outer capacity, III.32.

μ^*, outer measure, III.37.

I^+, outer capacity calculated using open sets, III.42.

D_A, debut of A, III.44.

$\mathfrak{m}_b^+(E)$, positive Radon measures on a completely regular space E, III.54.

$\mathcal{P}(E)$, space of Radon laws on E, III.60.

$\mathcal{F}_t$, $\mathcal{F}_{t+}$, $\mathcal{F}_{t-}$ (filtrations), IV.11.

∂, ζ, cemetery point and lifetime, IV.19.

r.c.l.l. etc, IV.16.

$f(t+)$, $f(t-)$, IV.20.

$U(f,I;[a,b])$, $D(f,I;[a,b])$, number of upcrossings and downcrossings, IV.21.

ess sup $f(s)$, ess lim sup $f(s)$, etc..., IV.36.

$\bar{f}(t) = \text{ess} \limsup_{s \downarrow\downarrow t} f(s)$, IV.37.

$\mathcal{F}_{\infty-}$, $\mathcal{F}_t (t < 0)$, $\mathcal{F}_{0-}$, $\mathcal{F}_\infty$, IV.47.

$\mathcal{F}_t^o$, IV.48.

$\mathcal{F}^o_T$, $\mathcal{F}^o_{T+}$, IV.52.

$S^{(n)}$, n-th dyadic approximation of S, IV.57.

$[\![U,V[\![$, etc..., stochastic intervals, IV.60.

$\mathcal{O}$, $\mathcal{P}$, optional, predictable σ-field, IV.61.

X_H, state of the process X at the instant H, IV.63.

$\mathcal{S}(T)$, sets of sequences of stopping times $\leq$ T, IV.79.

$A[(S_n)_{n \in \mathbb{N}}]$, set where (S_n) foretells S, IV.79.

BIBLIOGRAPHY

AZEMA, J. [1], Théorie générale des processus et retournement du temps, Annales E.N.S., 6, 1973, pp. 459-519.

BLACKWELL, D. [1], On a class of probability spaces, Proc. 3rd Berkeley Symposium, 2, pp. 16, University of Calif. Press, Berkeley, 1956.

BLACKWELL, D. and DUBINS, L.E. [1], A converse to the dominated convergence theorem, Ill. J. Math., 7, 1963, pp. 508-514.

BOURBAKI, N. [1], Eléments de mathématique. Espaces vectoriels topologiques, chapitre IV, 2nd edition, Act. Sci. et Industr. 1229, Hermann, Paris, 1964.

BOURBAKI, N. [2], Elements of Mathematics. General Topology, Chapter I-IV, Hermann Paris, 1966.

BOURBAKI, N. [3], Elements of mathematics. General Topology, Chapter V-X, Hermann Paris, 1966.

BOURBAKI, N. [4], Eléments de mathématique. Intégration, chapitre III, 2nd edition, Act. Sci. et Industr. 1175, Hermann, Paris, 1965.

BOURBAKI, N. [5], Eléments de mathématique. Intégration, chapitre IX, 2nd edition, Act. Sci. et Industr. 1343, Hermann, Paris 1969.

BRELOT, M. [1], Eléments de la théorie classique du potentiel, 3rd edition, Centre de Documentation Universitaire, Paris, 1965.

BRELOT, M. [2], Lectures on Potential theory, Tata Institute of Fund. Research, no. 19, Bombay, 1960.

CARLESON, L. [1], Selected problems on exceptional sets, Van Nostrand, New York, 1967.

CARTIER, P. [1], Processus aléatoires généralisés, Séminaire Bourbaki, 16-th year, 1963-64, exposé 272.

CHOQUET, G. [1], Theory of capacities, Annales Inst. Fourier, Grenoble, 5, 1953-54, pp. 131-295.

CHOQUET, G. [2], Forme abstraite du théorème de capacitabilité, Annales Inst. Fourier, Grenoble, 9, 1959, pp. 83-89.

CHOQUET, G. [3], Ensembles $\mathcal{K}$-analytiques et $\mathcal{K}$-sousliniens. Cas général et cas métrique, Annales Inst. Fourier, Grenoble, 9, 1959.

CHOU, C.S. and MEYER, P.A. [1], Sur la représentation des martingales comme intégrales stochastiques dans les processus ponctuels, Séminaire de Probabilités IX, Lecture Notes in Math. Springer, Berlin, 1975.

CHRISTENSEN, J.P.R. [1], Topology and Borel strutures, North Holland Mathematic Studies, no. 10, North Holland Publ. Co., Amsterdam, 1974.

CHUNG, K.L. [1], A course of probability theory, New York, Harcourt, Brace and World, 1968.

CHUNG, K.L. [2], On the fundamental hypotheses of Hunt processes, Instituto di alta matematica, Symposia Mathematica vo. IX, 1972, pp. 43-52.

CHUNG, K.L. and DOOB, J.L. [1], Fields, optionality and measurability, Amer. J. Math., 87, 1964, pp. 397-424.

CHUNG, K.L. and WALSH, J.B. [1], To reverse a Markov process, Acta Math., 123, 1969, pp. 225-251.

CORNEA, A. and LICEA, G. [1], Une démonstration unifiée des théorèmes de section de section de P.A. Meyer, Z.f. W-theorie, 10, 1968, pp. 198-202.

COURREGE, P. and PRIOURET, P. [1], Temps d'arrêt d'une fonction aléatoire, Publ. Inst. Stat. Univ. Paris, 14, 1965, pp. 245-274.

DELLACHERIE, C. [1], Capacités et processus stochastiques, Ergebn. der Math., no. 67, Springer, Berlin, 1972.

DELLACHERIE, C. [2], Ensembles analytiques : théorèmes de séparation et applications, Séminaire de Probabilités IX, Lecture Notes in Math., Springer, Berlin, 1975.

DELLACHERIE, C. [3], Les théorèmes de Mazurkiewicz-Sierpinski et Lusin, Séminaire de Probabilités V, Lecture Notes in Math. 191, Springer, Berlin, 1971.

DELLACHERIE, C. [4], Ensembles analytiques, capacités et mesures de Hausdorff, Lecture Notes in Math. 295, Springer, Berlin, 1972.

DELLACHERIE, C. [5], Un ensemble progressivement mesurable... Séminaire de Probabilités VIII, pp. 22-24. Lecture Notes in Math. 381, Springer, Berlin, 1974.

DELLACHERIE, C. [6], Sur les théorèmes fondamentaux de la théorie générale des processus, Séminaire de Probabilités VII, pp. 38-47, Lecture Notes in Math. 321, Springer, Berlin, 1973.

DELLACHERIE, C. and MEYER, P.A. [1], Ensembles analytiques et temps d'arrêt, Séminaire de Probabilités IX, Lecture Notes in Math., Springer, Berlin, 1975.

DELLACHERIE, C. and MEYER, P.A. [2], Un nouveau théorème de section et de projection, Séminaire de Probabilités IX, Lecture Notes in Math., Springer, Berlin, 1975.

DOLEANS-DADE, C. [1], Processus croissants naturels et processus croissants très-bien mesurables, C.R.A.S. 264, 1967.

DOLEANS-DADE, C. [2], Existence du processus croissant naturel associé à un potentiel de la classe (D), Z.f. W-theorie, 9, 1969, pp. 309-314.

DOOB, J.L. [1], Stochastic processes, New York, J. Wiley & Sons ; London, Chapman & Hall, 1953.

DOOB, J.L. [2], Brownian motion on a Green space, Theory of Prob. and its appl., 2, 1957, pp. 1-33.

DOOB, J.L. [3], Applications to analysis of a topological definition of smallness of a set, Bull. Amer. Math. Soc. 72, 1966, pp. 579-600.

DOOB, J.L. [4], Semimartingales and subharmonic functions, Trans. Amer. Math. Soc. 77, 1954, pp. 86-121.

DOOB, J.L. [5], A probability approach to the heat equation, Trans. Amer. Math. Soc. 80, 1955, pp 216-280.

DOOB, J.L. [6], Separability and measurable processes, J. Fac. Sci. Univ. of Tokyo Section 1A, 17, 1970, pp. 297-304.

DUNFORD, N. and SCHWARTZ, J.T. [1], Linear operators : Part I, General theory, New York, Interscience Publisher, 1963.

DYNKIN, E.B. [1], Markov processes, Grundlehren der math. Wiss. 121, Springer, Berlin, 1965 (Russian edition, 1962).

FERNIQUE, X. [1], Processus linéaires, processus généralisés, Ann. Inst. Fourier, Grenoble, 17-1, 1967, pp. 1-92.

GALMARINO, A.R. [1], Representation of an isotropic diffusion as a skew product, Z.f. W-theorie 1, 1963, pp. 359-378.

GELFAND, I.M. [1], Processus aléatoires généralisés, Dokl. Akad. Nauk SSSR 100, 1955, pp. 853-856.

HAUSDORFF, F. [1], Mengenlehre (3rd edition), Veit, Berlin, 1935, or Set Theory, Chelsea Publ. Co., New York, 1962.

HELMS, L.L. [1], Introduction to potential theory, J. Wiley & Son, New York, 1963.

HOFFMANN-JØRGENSEN, J. [1], The theory of analytic sets, Aarhus Univ. various publication series, no. 10, 1970.

HUNT, G.A. [1], Markoff processes and potentials 1, Illinois J. of Math. 1, 1957, pp. 44-93.

ITO, K. [1], The canonical modification of stochastic processes, J. Math. Soc. Japan 20, 1968, pp. 130-150.

ITO, K. [2], Canonical measurable random functions, Proc. Int. Conf. On Funct. Analysis, pp. 369-377, Univ. of Tokyo Press, 1970.

ITO, K. and McKEAN, H.P. [1], Diffusion processes and their sample paths, Grundlehren der math. Wiss. 125, Springer, Berlin, 1965.

KINGMAN, J.F.C. [1], Completely random measures, Pacific J. of Math. 21, 1967, pp. 59-78.

KNIGHT, F. [1], A predictive view of continuous time processes, Ann. of Prob., 1975.

KNIGHT, F. [2], Existence of small oscillations at zeros of Brownian motion, Séminaire de Probabilités VIII, pp. 134-149, Lecture Notes in Math. 381, Springer, Berlin, 1974.

KURATOWSKI, C. [1], Topologie I, 4th edition, Polska Akad. Nauk, Warszawa, 1958.

LE CAM, L. [1], Convergence in distribution of stochastic processes, Univ. Calif. Publ. in Statistics, no. 11, Berkeley, 1957.

LEVY, P. [1], Processus stochastiques et mouvement brownien, Gauthier-Villars, Paris, 1948.

Lusin, N.[1], Leçons sur les ensembles analytiques et leurs applications, Gauthier-Villars, Paris, 1930.

MAISONNEUVE, B. [1], Topologies du type de Skorokhod, Séminaire de Probabilités VI, pp. 113-117, Lecture Notes in Math. 258, Springer, Berlin, 1972.

MERTENS, J.F. [1], Théories des processus stochastiques généraux, applications aux surmartingales, Z.f. W-theorie 22, 1972, pp. 45-68.

MEYER, P.A. [1], Limites médiales d'après Mokobodzki, Séminaire de Probabilités VII, pp. 198-204, Lecture Notes in Math. 321, Springer, Berlin, 1973.

MEYER, P.A. [2], Temps d'arrêt algébriquement prévisibles, Séminaire de Probabilités VI, Lecture Notes in Math. 258, Springer, Berlin, 1972.

MEYER, P.A. [3], Une nouvelle démonstration des théorèmes de section, Séminaire de Probabilités III, pp. 155-159, Lecture Notes in Math., Springer, Berlin, 1969.

MEYER, P.A. [4], Une présentation de la théorie des ensembles sousliniens. Applications aux processus stochastiques, Séminaire Brelot-Choquet-Deny (théorie du potentiel), 7th year, 1962-63, 17 pages.

MEYER, P.A. [5], Guide détaillé de la théorie "générale" des processus, Séminaire de Probabilités II, pp. 140-165, Lecture Notes in Math. 51, Springer, Berlin, 1968.

MOKOBODZKI, G. [1], Ultrafiltres rapides sur $\mathbb{N}$, Séminaire Brelot-Choquet-Deny (théorie du potentiel), 12th year, 1967-68, no. 12, 22 pages (Institut Henri Poincaré, Paris).

MOKOBODZKI, G. [2], (Limites médiales : Séminaire Choquet.).

MORANDO, P. [1], Mesures aléatoires, Séminaire de Probabilités III, pp. 190-229, Lecture Notes in Math. 88, Springer, Berlin, 1969.

NEVEU, J. [1], Bases mathématiques du calcul des probabilités, Masson, Paris, 1964.

PARTHASARATHY, K.R. [1], Probability measures on metric spaces, Academic Press, New York, 1967.

PFANZAGL, J. and PIERLO, W. [1], Compact systems of sets, Lecture Notes in Math. 16, Springer, Berlin, 1966.

PREISS, D. [1], Metric spaces in which Prokhorov's theorem is not valid, Z.f. W-theorie 27, 1973, pp. 109-116.

PROKHOROV, Y.V. [1], Convergence of random processes and limit theorems in probability theory, Th. Prob. and its Appl. 1, 1956, pp. 157-214.

ROGERS, C.A. [1], Lusin's second theorem of separation, J. London Math. Soc. 6, 1973, pp. 491-503.

SAKS, S. [1], Theory of the integral, translated by L.C. Young, 2nd edition, Hafner Publ. Co., New York, 1937.

SION M. [1], On capacitability and measurability, Ann. Inst. Fourier, Grenoble, 13, 1963, pp. 88-99.

SION, M. [2], On analytic sets in topological spaces, Trans. Amer. Math. Soc. 96, 1960, pp. 341-354.

SION, M. [3], Topological and measure theoretic properties of analytic sets, Proc. Amer. Math. Soc. 11, 1960, pp. 769-776.

SION, M. [4], On uniformization of sets in topological spaces, Trans. Amer. Math. Soc. 96, 1960, pp. 237-246.

SION, M. [5], Continuous images of Borel sets, Proc. Amer. Math. Soc. 12, 1961, pp. 385-391.

SION, M. and BRESSLER, D.W. [1], The current theory of analytic sets, Can. J. of Math. 16, 1964, pp. 207-230.

STRASSEN, V. [1], The existence of probability measures with given marginals, Ann. Math. Stat. 36, 1965, pp. 88-91.

STROOK, D. [1], The Kac approach to potential theory, J. Math. Mech. 16, 1967, pp. 829-852.

WALSH, J.B. [1], Some topologies connected with Lebesgue measure, Séminaire de Probabilités V, 1971, pp. 290-310, Lecture Notes in Math. 191, Springer, Berlin, 1971.

WALSH, J.B. [2], The perfection of multiplicative functionals, Séminaire de Probabilités VI, pp. 233-252, Lecture Notes in Math. 258, Springer, Berlin, 1972.

YEN, K.A. [1], Forme mesurable de la théorie des ensembles sousliniens, applications à la théorie de la mesure, Scientia Sinica, 1975.